A Short Course in Matrix Theory

A Short Course in Matrix Theory

Erwin and Margaret Kleinfeld

NOVA SCIENCE PUBLISHERS, INC.
COMMACK, NEW YORK

Art Director: Maria Ester Hawrys
Assistant Director: Elenor Kallberg
Graphics: Susan Boriotti and Frank Grucci
Manuscript Coordinator: Phyllis Gaynor
Book Production: Gavin Aghamore, Joanne Bennette, Michelle Keller
Christine Mathosian and Tammy Sauter
Circulation: Iyatunde Abdullah, Cathy DeGregory and Annette Hellinger

*Library of Congress Cataloging-in-Publication Data
available upon request*

ISBN 1-56072-422-6

Copyright © 1997 by Nova Science Publishers, Inc.
6080 Jericho Turnpike, Suite 207
Commack, New York 11725
Tele. 516-499-3103   Fax 516-499-3146
E Mail Novascil@aol.com
Novascience@earthlink.net

All rights reserved. No part of this book may be reproduced, stored in a retrieval system or transmitted in any form or by any means: electronic, electrostatic, magnetic, tape, mechanical photocopying, recording or otherwise without permission from the publishers

Printed in the United States of America

Contents

Introduction 1

1 Vectors and Matrices   **3**
    Section 1.1 Matrix Arithmetic 3
    Section 1.2 Vectors 7
    Section 1.3 Reduced Row Echelon Form 12
    Section 1.4 Inverse 21
    Section 1.5 Rank and Dimension 27
    Section 1.6 Null Space and Hermite Form 33

2 Determinants   **39**
    Section 2.1 Definition and Properties 39
    Section 2.2 Applications of Determinant 47

3 Diagonalization   **51**
    Section 3.1 Eigenvalues and Eigenvectors 51
    Section 3.2 Diagonalization 59
    Section 3.3 Orthogonal Basis 66
    Section 3.4 Orthogonal Diagonalization 73

4 Applications   **83**
    Section 4.1 Linear Equations 83
    Section 4.2 Differential Equations 91
    Section 4.3 Powers of a Matrix and the Fibonacci Sequence . . 96
    Section 4.4 Least Squares Approximation 99

5 Geometry   **103**
    Section 5.1 Cross Product and Planes in R^3 103
    Section 5.2 Lines in R^3 107

Section 5.3 Projection 113
Section 5.4 Orthogonal Matrices and Quadrics 115

A Change of Basis   **123**

B Answers to Selected Exercises   **129**

C MATS   **149**

Index   **157**

Introduction

In this book we attempt to give the student the essentials of matrix theory and linear algebra as quickly and painlessly as possible. Concepts and techniques are introduced by example. Proofs are included where they are enlightening, and omitted when they are distracting. The goal is to reach the topics of eigenvalues, eigenvectors, diagonalization and some applications within the time constraints of a short course. This goal gives a unity and flow to the book that is often missing in fatter books on the subject. The book is self-contained and does not require calculus as a prerequisite. We use a matrix approach and work within the vector space R^n. The student is introduced to mathematical thinking without the intimidating theorem-proof terminology.

Full advantage is taken of the power of the row reduced echelon form. The Hermite form is reintroduced and used to develop a convenient algorithm for finding a basis of the null space of a matrix. Included with this book is a computer program called MATS, designed by our colleague Prof. Eugene Johnson especially for use by students taking a first course in linear algebra. A student can learn to use MATS in a few minutes. MATS can do elementary row operations, compute the reduced row echelon form (RREF) and Hermite form of a matrix, do basic matrix operations and compute determinants, inverses, etc. MATS was designed as an educational tool, and can take the computational drudgery out of the course, while still allowing the student to see the algorithms at work.

Chapter 1 introduces the student to matrix arithmetic, vectors and vector operations in R^n; the concepts of subspace, basis, and dimension in R^n; the row space, column space and null space of a matrix, as well as the RREF, Hermite form, and inverse of a matrix. By restricting to R^n, and omitting most proofs, all this can be done more quickly and simply than is usual. Students frequently say that examples give them a better understanding of the concepts than proofs, and we have tried to take advantage of this.

In Chapter 2 we define determinants for 1×1 and 2×2 matrices, and for larger size matrices by the expansion by minors formula. We quickly explain the basic properties of determinants, some methods for computing them, and a couple of applications (Cramer's Rule and Vandermonde Determinants).

By Chapter 3 we are ready to compute eigenvalues and eigenvectors and to diagonalize a matrix. We then discuss orthogonality and the orthogonal diagonalization of symmetric matrices. This chapter is the heart of the course, using everything that has gone before.

Chapter 4 applies what has been learned to systems of linear equations, systems of first order linear differential equations, powers of a matrix, and least squares approximations.

In Chapter 5 we apply what has been learned to some geometry in R^2 and R^3. We include three appendices. In Appendix A (Change of Basis) we show how to compute the coordinates of a vector in R^n with respect to an arbitrary basis and how the coordinates of the same vector with respect to different bases are related. We also define the matrix of a linear transformation with respect to an arbitrary basis and show how it is related to the standard matrix of the same linear transformation. Appendix B contains answers to selected exercises. Of course some of the answers are not unique, so that the answer given may not be the only correct answer possible. In Appendix C, we discuss the MATS program and give some instructions for its use and a list of commands.

A more detailed and rigorous treatment of the material in this book with elementary proofs of the main theorems may be found in the book "Elementary Linear Algebra" by the same authors published by Nova Science Publishers, Inc. 1996.

Chapter 1

Vectors and Matrices

Section 1.1 Matrix Arithmetic

A matrix is a rectangular array of numbers, for example

$$A = \begin{bmatrix} 1 & 3 & -2 \\ 0 & 2 & 2 \\ 5 & -\frac{1}{3} & 0 \\ 10 & 4 & -6 \end{bmatrix}.$$

The matrix A above is a 4×3 matrix, meaning it has 4 rows and 3 columns. Two matrices are equal only if they are identical (same size and same numbers in every position). We can refer to individual numbers in the matrix by giving the row and column. For example in the matrix A above -6 occurs in the 4th row and 3rd column. Two matrices of the same size can be added by adding corresponding entries, for example

$$\begin{bmatrix} 1 & 3 & -2 \\ 0 & 2 & 2 \\ 5 & -\frac{1}{3} & 0 \\ 10 & 4 & -6 \end{bmatrix} + \begin{bmatrix} 3 & 2 & -1 \\ 1 & 0 & \frac{1}{5} \\ 2 & \frac{2}{3} & 3 \\ 4 & 3 & -2 \end{bmatrix} = \begin{bmatrix} 4 & 5 & -3 \\ 1 & 2 & \frac{11}{5} \\ 7 & \frac{1}{3} & 3 \\ 14 & 7 & -8 \end{bmatrix}.$$

The product of two matrices A and B is denoted by $A \cdot B$ or AB. The number of columns of A must equal the number of rows of B or else the product is not defined. For example

$$\begin{bmatrix} 1 & -5 & 3 \end{bmatrix} \begin{bmatrix} 6 \\ 1 \\ 1 \end{bmatrix} = \begin{bmatrix} 1 \cdot 6 + (-5) \cdot 1 + 3 \cdot 1 \end{bmatrix} = \begin{bmatrix} 4 \end{bmatrix},$$

$$\begin{bmatrix} 1 & -5 & 3 \\ 2 & 0 & -1 \\ 3 & 1 & 0 \end{bmatrix} \begin{bmatrix} 6 \\ 1 \\ 1 \end{bmatrix} = \begin{bmatrix} 1 \cdot 6 + (-5) \cdot 1 + 3 \cdot 1 \\ 2 \cdot 6 + 0 \cdot 1 + (-1) \cdot 1 \\ 3 \cdot 6 + 1 \cdot 1 + 0 \cdot 1 \end{bmatrix} = \begin{bmatrix} 4 \\ 11 \\ 19 \end{bmatrix},$$

and

$$\begin{bmatrix} 1 & -5 & 3 \\ 2 & 0 & -1 \\ 3 & 1 & 0 \end{bmatrix} \begin{bmatrix} 6 & 0 & 1 \\ 1 & 1 & 0 \\ 1 & 0 & 1 \end{bmatrix} = \begin{bmatrix} 4 & -5 & 4 \\ 11 & 0 & 1 \\ 19 & 1 & 3 \end{bmatrix}.$$

The product has as many rows as the matrix on the left, and as many columns as the matrix on the right. The entry in the second row and third column of the product can be obtained by calculating the product of the second row of the matrix on the left times the third column of the matrix on the right.

$$\begin{bmatrix} 2 & 0 & -1 \end{bmatrix} \begin{bmatrix} 1 \\ 0 \\ 1 \end{bmatrix} = \begin{bmatrix} 1 \end{bmatrix}.$$

One of the surprising facts about matrix multiplication is

$$(AB)C = A(BC)$$

whenever multiplication is possible. This rule is called the associative law of matrix multiplication. We also have

$$A(B + C) = AB + AC,$$

and

$$(A + B)C = AC + BC,$$

called the distributive laws. However, for matrix multiplication AB need not equal BA, matrix multiplication is not commutative. For example

$$\begin{bmatrix} 0 & 1 \\ 0 & 0 \end{bmatrix} \begin{bmatrix} 0 & 0 \\ 0 & 1 \end{bmatrix} = \begin{bmatrix} 0 & 1 \\ 0 & 0 \end{bmatrix},$$

Vectors and Matrices

while
$$\begin{bmatrix} 0 & 0 \\ 0 & 1 \end{bmatrix} \begin{bmatrix} 0 & 1 \\ 0 & 0 \end{bmatrix} = \begin{bmatrix} 0 & 0 \\ 0 & 0 \end{bmatrix}.$$

The zero matrix of any size is the matrix of that size with all entries equal to zero. We will use $[0]$ to stand for the zero matrix of any size, where the size can be determined by the context. The above example shows that we can have $AB = [0]$ while $A \neq [0]$ and $B \neq [0]$. Since $A[0] = [0]$, the above example also shows that we cannot cancel the A, for $AB = A[0]$, but $B \neq [0]$. So we see that many but not all of the rules that apply to the arithmetic of numbers are also valid in matrix arithmetic.

The matrix $\begin{bmatrix} 1 & 0 \\ 0 & 1 \end{bmatrix}$ is called the 2×2 identity matrix. The matrix $\begin{bmatrix} 1 & 0 & 0 \\ 0 & 1 & 0 \\ 0 & 0 & 1 \end{bmatrix}$ is called the 3×3 identity matrix. For each positive integer n, the square matrix of size $n \times n$ with 1 on the main (left to right) diagonal and 0 elsewhere is the identity matrix of size n. It will be denoted by I. For an arbitrary $n \times n$ matrix A we have

$$AI = IA = A.$$

In the arithmetic of $n \times n$ matrices, the identity matrix acts like the number 1, and the zero matrix acts like the number 0.

Exercises

1. Calculate the following matrix sums:

   (a) $\begin{bmatrix} 1 & 2 \\ 3 & 4 \end{bmatrix} + \begin{bmatrix} 4 & 3 \\ 2 & 1 \end{bmatrix}$,

   (b) $\begin{bmatrix} 2 & 0 \\ 5 & -1 \end{bmatrix} + \begin{bmatrix} -2 & 0 \\ -5 & 1 \end{bmatrix}$,

   (c) $\begin{bmatrix} 0 & 1 \\ 1 & 0 \end{bmatrix} + \begin{bmatrix} 1 & 0 \\ 0 & 1 \end{bmatrix}$,

   (d) $\begin{bmatrix} 1 & 0 & 0 \\ 0 & 2 & 0 \\ 0 & 0 & 3 \end{bmatrix} + \begin{bmatrix} 0 & 0 & 1 \\ 0 & 2 & 0 \\ 3 & 0 & 0 \end{bmatrix}$.

2. Calculate the following matrix products where possible:

(a) $\begin{bmatrix} 3 & 5 & 1 \\ 2 & -1 & 3 \end{bmatrix} \begin{bmatrix} 1 & -1 \\ 2 & 3 \\ -4 & 2 \end{bmatrix}$

(b) $\begin{bmatrix} 3 & 5 & 1 \\ 2 & -1 & 3 \end{bmatrix} \begin{bmatrix} 1 & -1 & 1 \\ 2 & 3 & 0 \\ -4 & 2 & 1 \end{bmatrix}$

(c) $\begin{bmatrix} 1 & -1 \\ 2 & 3 \\ -4 & 2 \end{bmatrix} \begin{bmatrix} 3 & 5 & 1 \\ 2 & -1 & 3 \end{bmatrix}$

(d) $\begin{bmatrix} 1 & -1 & 1 \\ 2 & 3 & 0 \\ -4 & 2 & 1 \end{bmatrix} \begin{bmatrix} 3 & 5 & 1 \\ 2 & -1 & 3 \end{bmatrix}$.

3. Calculate the following matrix products:

(a) $\begin{bmatrix} 0 & 1 \\ 1 & 0 \end{bmatrix} \begin{bmatrix} 1 & 0 \\ 0 & 1 \end{bmatrix}$

(b) $\begin{bmatrix} 0 & 1 \\ 0 & 0 \end{bmatrix} \begin{bmatrix} 0 & 0 \\ 1 & 0 \end{bmatrix}$,

(c) $\begin{bmatrix} 0 & 0 \\ 1 & 0 \end{bmatrix} \begin{bmatrix} 0 & 1 \\ 0 & 0 \end{bmatrix}$,

(d) $\begin{bmatrix} 0 & 0 & 0 \\ 0 & 0 & 0 \\ 0 & 0 & 0 \end{bmatrix} \begin{bmatrix} -1 & 5 & -3 \\ 2 & 2 & 9 \\ 8 & 6 & 4 \end{bmatrix}$,

(e) $\begin{bmatrix} 0 & 1 & 0 \\ 1 & 0 & 0 \\ 0 & 0 & 1 \end{bmatrix} \begin{bmatrix} 9 & 8 & 7 \\ 6 & 5 & 4 \\ 3 & 2 & 1 \end{bmatrix}$,

(f) $\begin{bmatrix} 9 & 8 & 7 \\ 6 & 5 & 4 \\ 3 & 2 & 1 \end{bmatrix} \begin{bmatrix} 0 & 1 & 0 \\ 1 & 0 & 0 \\ 0 & 0 & 1 \end{bmatrix}$,

(g) $\begin{bmatrix} 2 & 0 & 0 \\ 0 & 2 & 0 \\ 0 & 0 & 2 \end{bmatrix} \begin{bmatrix} 1 & 2 & 3 \\ 4 & 5 & 6 \\ 7 & 8 & 9 \end{bmatrix}$.

Vectors and Matrices

4. Calculate the following matrix products:

   (a) $\begin{bmatrix} c \end{bmatrix} \begin{bmatrix} 1 & 2 & 3 & 4 & 5 & 6 \end{bmatrix}$

   (b) $\begin{bmatrix} c & d \end{bmatrix} \begin{bmatrix} 1 & 0 & 2 & 0 & 3 \\ 0 & 4 & 0 & 5 & 0 \end{bmatrix}$,

   (c) $\begin{bmatrix} e & f \end{bmatrix} \begin{bmatrix} 1 & 2 & 3 & 4 \\ 5 & 6 & 7 & 8 \end{bmatrix}$,

   (d) $\begin{bmatrix} 1 & 2 & 3 & 4 \end{bmatrix} \begin{bmatrix} 5 \\ 6 \\ 7 \\ 8 \end{bmatrix}$,

   (e) $\begin{bmatrix} 5 \\ 6 \\ 7 \\ 8 \end{bmatrix} \begin{bmatrix} 1 & 2 & 3 & 4 \end{bmatrix}$,

   (f) $\begin{bmatrix} 1 & 1 & 1 \\ 0 & 1 & 0 \\ 0 & 0 & 1 \end{bmatrix} \begin{bmatrix} 4 & 1 & 1 \\ 1 & 4 & 1 \\ 1 & 1 & 4 \end{bmatrix}$.

5. Show that there exists no 2×2 matrix B satisfying the matrix equation
$$\begin{bmatrix} 1 & 2 \\ 2 & 4 \end{bmatrix} B = \begin{bmatrix} 1 & 0 \\ 0 & 1 \end{bmatrix}.$$

Section 1.2 Vectors

A **vector** in R^n is an ordered set of n numbers. One can think of it as the entries of a particular row of an $n \times n$ matrix M. In that case we call the vector a row vector. For example let

$$M = \begin{bmatrix} 1 & 2 & 5 \\ 3 & -1 & 6 \\ 8 & 0 & 7 \end{bmatrix}.$$

The first row of M can be considered a vector in R^3. We could write this vector as $\begin{bmatrix} 1 & 2 & 5 \end{bmatrix}$, a 1×3 matrix, but we will usually write $(1, 2, 5)$. The columns of M can also be considered vectors in R^3. The column vectors of M are

$$\begin{bmatrix} 1 \\ 3 \\ 8 \end{bmatrix}, \begin{bmatrix} 2 \\ -1 \\ 0 \end{bmatrix}, \text{ and } \begin{bmatrix} 5 \\ 6 \\ 7 \end{bmatrix}.$$

When considering them as vectors in R^n they can also be written as $(1, 3, 8)$, $(2, -1, 0)$, and $(5, 6, 7)$.

The **addition of two vectors** is like the addition of two $1 \times n$ (or $n \times 1$) matrices. For example

$$(1, 3, 8) + (2, -1, 0) = (3, 2, 8),$$

and

$$(3, -1, 6) + (8, 0, 7) = (11, -1, 13).$$

Vector addition is both associative and commutative. We use **O** to denote the **zero vector** in R^n. In R^3, $\mathbf{O} = (0, 0, 0)$.

The **additive inverse** of the vector $(1, 2, -3, 4)$ in R^4 is $(-1, -2, 3, -4)$. Note that $(1, 2, -3, 4) + (-1, -2, 3, -4) = \mathbf{O} = (0, 0, 0, 0)$.

Numbers are sometimes called **scalars** to distinguish them from vectors. When we multiply a vector by a number we call this **scalar multiplication**. To multiply a vector by a number c, simply multiply every entry (component) of the vector by c. For example

$$c(1, 2, 3, 4) = (c, 2c, 3c, 4c).$$

In matrix form this could be written

$$\begin{bmatrix} c \end{bmatrix} \begin{bmatrix} 1 & 2 & 3 & 4 \end{bmatrix} = \begin{bmatrix} c & 2c & 3c & 4c \end{bmatrix}.$$

Thus we have

$$5(2, -1, 3, 3) = (10, -5, 15, 15).$$

The idea of a **linear combination** of vectors is very important. If **u** and **v** are any two vectors in R^n and c and d are any two numbers, then $c\mathbf{u}+d\mathbf{v}$ is a linear combination of $\mathbf{u}$ and $\mathbf{v}$. If there are k vectors we use subscripts and denote the k vectors by $\mathbf{v}_1, \mathbf{v}_2, \cdots \mathbf{v}_k$. Then if we also have k numbers $c_1, c_2, \cdots, c_k$, we can form $c_1\mathbf{v}_1 + c_2\mathbf{v}_2 + \cdots + c_k\mathbf{v}_k$. This is a linear combination of the vectors $\mathbf{v}_1, \mathbf{v}_2, \cdots \mathbf{v}_k$.

For example, let's go back to the row vectors of the matrix M, $(1, 2, 5)$, $(3, -1, 6)$, and $(8, 0, 7)$. Then

$$\begin{aligned} & 2(1,2,5) + 4(3,-1,6) + (-1)(8,0,7) \\ = & (2,4,10) + (12,-4,24) + (-8,0,-7) \\ = & (6,0,27). \end{aligned}$$

This shows that $(6, 0, 27)$ is a linear combination of the row vectors of M.

Example $(1, 1, 1)$ *is not a linear combination of* $(1, -1, 0)$, $(2, 3, 0)$, *and* $(4, 1, 0)$.

Reason. All linear combinations of those three vectors must have the last entry equal to 0.

A **subspace** of R^n is defined as any set W of vectors "closed" under linear combination. This means that a set W is a subspace of R^n if all linear combinations of vectors in W are also in W. In other words you cannot get outside a subspace W by taking linear combinations of vectors that are in W.

Example *The set of all vectors of the form* (c, c, c), *all entries equal, is a subspace of* R^3.

Reason. $(c, c, c) + (d, d, d) = (c + d, c + d, c + d)$ and $k(c, c, c) = (kc, kc, kc)$. Thus the set is closed under vector addition and scalar multiplication. This is enough to obtain closure under linear combinations.

Example *The set of all vectors of the form* $(c, 0, d)$, *second entry 0, is a subspace of* R^3.

Reason. $(c,0,d)+(e,0,f) = (c+e,0,d+f)$ and $k(c,0,d) = (kc,0,kd)$.

Except for the set containing only the zero vector, all other subspaces of R^n must contain infinitely many vectors.

The **span** of a set of vectors $S = \{v_1, v_2, \cdots v_k\}$ is the set of all linear combinations of the vectors $v_1, v_2, \cdots v_k$. This is always a subspace, for any set S. If the span of $v_1, v_2, \cdots v_k$ is all of R^n, we say that the set S **spans** R^n.

Example *The set $\{(1,1),(2,3),(3,2)\}$ spans R^2.*

Reason. In fact any two of the three vectors in this set span R^2. For instance using $(1,1)$ and $(2,3)$ we have $(1,0) = 3(1,1) - (2,3)$, and $(0,1) = (2,3) - 2(1,1)$, so that $(x,0) = 3x(1,1) - x(2,3)$ and $(0,y) = y(2,3) - 2y(1,1)$, from which it follows that

$$(x,y) = (x,0) + (0,y) = (3x-2y)(1,1) + (y-x)(2,3).$$

For the sake of efficiency we seek a smallest set of vectors to span a given subspace of R^n. For this purpose we introduce the term "linearly independent". A set $S = \{v_1, v_2, \cdots v_k\}$ of vectors is **linearly independent** if $c_1 v_1 + c_2 v_2 + \cdots + c_k v_k = O$ only if $c_i = 0$ for all $i = 1, 2, \cdots, k$. In other words the only way to get the zero vector as a linear combination of a linearly independent set of vectors is to use all zeros as the coefficients.

Example *The set $\{(1,1),(2,3),(3,2)\}$ is not linearly independent.*

Reason. From the previous example we have $(3,2) = 5(1,1) - (2,3)$, using $x = 3$ and $y = 2$. Thus $(3,2) - 5(1,1) + (2,3) = (0,0)$.

A set is **linearly dependent** whenever it is not linearly independent.

Example *Any set $S = \{O, v_1, v_2, \cdots v_k\}$ which includes the zero vector is linearly dependent.*

Vectors and Matrices

Reason. $1\mathbf{O} + 0\mathbf{v}_1 + 0\mathbf{v}_2 + \cdots + 0\mathbf{v}_k = \mathbf{O}$.

Adjoin any vectors to a set which is linearly dependent and the new set is also linearly dependent. Delete a vector from a linearly dependent set and the new set may or may not be linearly dependent. For linearly independent sets it is just the opposite: delete any vector and the new set is linearly independent, but add vectors to the set and the new set may or may not be linearly independent.

A **basis** of a subspace W of R^n is a set S which spans W and is also a linearly independent set.

Exercises

1. Determine whether or not the following sets of vectors span R^2:

    (a) $\{(1,2),(2,3),(3,4)\}$
    (b) $\{(1,2),(2,4)\}$
    (c) $\{(1,2),(2,3)\}$.

2. Determine whether or not the following sets of vectors span R^3:

    (a) $\{(1,2,5),(3,-1,6),(8,0,7)\}$
    (b) $\{(1,0,0),(0,1,0),(0,0,1)\}$
    (c) $\{(1,3,8),(2,-1,0),(5,6,7)\}$
    (d) $\{(1,0,2),(2,3,4),(-1,1,-2)\}$
    (e) $\{(1,2,-1),(0,3,1),(2,4,-2)\}$
    (f) $\{(1,2,5),(1,3,1)\}$.

3. Determine whether or not the vectors in exercise 1 form a basis of R^2.

4. Determine whether or not the vectors in exercise 2 form a basis of R^3.

5. For each of the following matrices, form the sets S_1 of the row vectors and S_2 of the column vectors and in each case determine whether or not the sets are a basis of R^3.

(a) $\begin{bmatrix} 1 & 1 & 1 \\ 0 & 1 & 1 \\ 0 & 1 & 2 \end{bmatrix}$

(b) $\begin{bmatrix} 1 & 2 & 3 \\ 4 & 5 & 6 \\ 7 & 8 & 9 \end{bmatrix}$

(c) $\begin{bmatrix} 1 & 1 & 0 \\ 1 & 1 & 0 \\ 0 & 0 & 1 \end{bmatrix}$

(d) $\begin{bmatrix} 1 & 2 & 3 \\ 0 & 1 & 2 \\ 0 & 0 & 1 \end{bmatrix}$.

Section 1.3 Reduced Row Echelon Form

The matrix equation

$$\begin{bmatrix} 1 & 2 & -1 \\ 2 & 3 & 4 \\ 1 & 0 & 1 \end{bmatrix} \begin{bmatrix} x \\ y \\ z \end{bmatrix} = \begin{bmatrix} 1 \\ 1 \\ 2 \end{bmatrix}$$

in the unknowns x, y, and z may be written as

$$\begin{aligned} x + 2y - z &= 1 \\ 2x + 3y + 4z &= 1 \\ x \phantom{{}+2y} + z &= 2 \end{aligned}$$

and so represents a system of 3 linear equations in 3 unknowns. The word "linear" means that the unknowns may be multiplied by numbers and added but no other operations such as squaring or multiplying two unknowns. (For example the equation $xy + 3z = 0$ is not linear.)

We are interested in solving such a system. Sometimes we will start with the system of equations and then write it in matrix form.

Vectors and Matrices

A second way of looking at this system of equations is as the vector equation

$$x \begin{bmatrix} 1 \\ 2 \\ 1 \end{bmatrix} + y \begin{bmatrix} 2 \\ 3 \\ 0 \end{bmatrix} + z \begin{bmatrix} -1 \\ 4 \\ 1 \end{bmatrix} = \begin{bmatrix} 1 \\ 1 \\ 2 \end{bmatrix}.$$

In this form we are asking how to write the column vector $\begin{bmatrix} 1 \\ 1 \\ 2 \end{bmatrix}$ as a linear combination of the column vectors of the given matrix. Both ways of viewing the matrix equation are very important.

We now introduce a method that will permit us to solve such a system by operating on a matrix. The matrix we will operate on is called the augmented matrix of the system. Simply adjoin the column vector on the right hand side of the equation as an additional column to the given (coefficient) matrix. In the system above the coefficient matrix is

$$\begin{bmatrix} 1 & 2 & -1 \\ 2 & 3 & 4 \\ 1 & 0 & 1 \end{bmatrix},$$

and the augmented matrix is

$$\begin{bmatrix} 1 & 2 & -1 & 1 \\ 2 & 3 & 4 & 1 \\ 1 & 0 & 1 & 2 \end{bmatrix}.$$

We will operate on the augmented matrix of the system using three special types of operations that do not change the solutions of the system. These operations are called

Elementary Row Operations

Type 1. Add a multiple of one row vector to another.

Type 2. Multiply a row vector by a nonzero number.

Type 3. Exchange (switch) two row vectors.

Two matrices are called row equivalent if one can be changed into the other finite by a sequence of elementary row operations.

The object is to perform a sequence of elementary row operations on the augmented matrix of the system until we obtain a matrix of a simple form from which the solutions can be easily determined. The simple form we are aiming for is called the reduced row echelon form (RREF).

Reduced Row Echelon Form

A matrix is in RREF if it satisfies the following four conditions:

1. The first nonzero entry in each row is a 1 (called the leading 1 of that row.)

2. The leading 1 in row $j+1$ is to the right of the leading 1 in row j. (The leading 1s move from left to right as you go down.)

3. A column which contains the leading 1 of some row must have zeros everywhere else in the column.

4. The zero rows if any come last.

Every matrix A is row equivalent to one and only one matrix in reduced row echelon form. This matrix is called $RREF(A)$, the reduced row echelon form of A.

Different matrices may have the same RREF. In fact if two matrices are row equivalent they will have the same RREF, however if you start with a given matrix A and proceed by elementary row operations to a reduced row echelon form matrix, the resulting matrix is unique. There is only one correct RREF of A, and there always is one. This is the content of the above statement.

$RREF(A)$ is a very useful tool and gives a great deal of information about the matrix A and the system of equations it represents. Finding $RREF(A)$ can be tedious. There are computer programs that will produce it for you. It is also a good idea to learn how to do this without a computer. You can use the computer as a check.

Vectors and Matrices

Example *The following is a sequence of matrices which starts with the augmented matrix of the system of equations introduced at the beginning of this section and ends with the RREF of this matrix.*

$$\begin{bmatrix} 1 & 2 & -1 & 1 \\ 2 & 3 & 4 & 1 \\ 1 & 0 & 1 & 2 \end{bmatrix} \to \begin{bmatrix} 1 & 2 & -1 & 1 \\ 0 & -1 & 6 & -1 \\ 1 & 0 & 1 & 2 \end{bmatrix} \to \begin{bmatrix} 1 & 2 & -1 & 1 \\ 0 & -1 & 6 & -1 \\ 0 & -2 & 2 & 1 \end{bmatrix}$$

$$\to \begin{bmatrix} 1 & 2 & -1 & 1 \\ 0 & 1 & -6 & 1 \\ 0 & -2 & 2 & 1 \end{bmatrix} \to \begin{bmatrix} 1 & 0 & 11 & -1 \\ 0 & 1 & -6 & 1 \\ 0 & -2 & 2 & 1 \end{bmatrix} \to \begin{bmatrix} 1 & 0 & 11 & -1 \\ 0 & 1 & -6 & 1 \\ 0 & 0 & -10 & 3 \end{bmatrix}$$

$$\to \begin{bmatrix} 1 & 0 & 11 & -1 \\ 0 & 1 & -6 & 1 \\ 0 & 0 & 1 & -\frac{3}{10} \end{bmatrix} \to \begin{bmatrix} 1 & 0 & 0 & \frac{23}{10} \\ 0 & 1 & -6 & 1 \\ 0 & 0 & 1 & -\frac{3}{10} \end{bmatrix} \to \begin{bmatrix} 1 & 0 & 0 & \frac{23}{10} \\ 0 & 1 & 0 & -\frac{8}{10} \\ 0 & 0 & 1 & -\frac{3}{10} \end{bmatrix}.$$

Each matrix comes from the preceding one by just one elementary row operation. You should be able to deduce what row operation was used each time. The resulting RREF matrix represents the system of equations

$$\begin{array}{rcrcrcl} 1x & + & 0y & + & 0z & = & \frac{23}{10} \\ 0x & + & 1y & + & 0z & = & -\frac{8}{10} \\ 0x & + & 0y & + & 1z & = & -\frac{3}{10} \end{array},$$

to which there is the unique and obvious solution $x = 23/10$, $y = -8/10$, *and* $z = -3/10$. *This also tells us how to write the column vector* $\begin{bmatrix} 1 \\ 1 \\ 2 \end{bmatrix}$ *as a linear combination of the columns of the matrix*

$$\begin{bmatrix} 1 & 2 & -1 \\ 2 & 3 & 4 \\ 1 & 0 & 1 \end{bmatrix},$$

namely

$$\begin{bmatrix} 1 \\ 1 \\ 2 \end{bmatrix} = 23/10 \begin{bmatrix} 1 \\ 2 \\ 1 \end{bmatrix} + (-8/10) \begin{bmatrix} 2 \\ 3 \\ 0 \end{bmatrix} + (-3/10) \begin{bmatrix} -1 \\ 4 \\ 1 \end{bmatrix}.$$

You can check that this is correct. If we had started with the matrix

$$\begin{bmatrix} 1 & 2 & -1 & 0 \\ 2 & 3 & 4 & 0 \\ 1 & 0 & 1 & 0 \end{bmatrix}$$

and done the same row operations, we would have arrived at the RREF form matrix

$$\begin{bmatrix} 1 & 0 & 0 & 0 \\ 0 & 1 & 0 & 0 \\ 0 & 0 & 1 & 0 \end{bmatrix},$$

which shows that the system $\begin{bmatrix} 1 & 2 & -1 \\ 2 & 3 & 4 \\ 1 & 0 & 1 \end{bmatrix} \begin{bmatrix} x \\ y \\ z \end{bmatrix} = \begin{bmatrix} 0 \\ 0 \\ 0 \end{bmatrix}$ having the same solutions as the system

$$\begin{bmatrix} 1 & 0 & 0 \\ 0 & 1 & 0 \\ 0 & 0 & 1 \end{bmatrix} \begin{bmatrix} x \\ y \\ z \end{bmatrix} = \begin{bmatrix} 0 \\ 0 \\ 0 \end{bmatrix},$$

has only the zero solution $x = 0$, $y = 0$, and $z = 0$. This tells us that the column vectors

$$\begin{bmatrix} 1 \\ 2 \\ 1 \end{bmatrix}, \begin{bmatrix} 2 \\ 3 \\ 0 \end{bmatrix}, \begin{bmatrix} -1 \\ 4 \\ 1 \end{bmatrix}$$

are linearly independent since

$$x \begin{bmatrix} 1 \\ 2 \\ 1 \end{bmatrix} + y \begin{bmatrix} 2 \\ 3 \\ 0 \end{bmatrix} + z \begin{bmatrix} -1 \\ 4 \\ 1 \end{bmatrix} = \begin{bmatrix} 0 \\ 0 \\ 0 \end{bmatrix}$$

has only the zero solution.

Procedure To determine if a set of vectors in R^n is linearly independent, put the vectors as the columns of a matrix and find the RREF. If there is a leading one in every column of the RREF, the vectors you started with are a linearly independent set. If there is not a leading one in every column of the RREF, the vectors you started with are not a linearly independent set.

Vectors and Matrices

Let A be an $n \times n$ matrix and suppose that $RREF(A)$ is

$$RREF(A) = \begin{bmatrix} 1 & 2 & 0 & 5 \\ 0 & 0 & 1 & 3 \\ 0 & 0 & 0 & 0 \\ 0 & 0 & 0 & 0 \end{bmatrix}.$$

Let $\mathbf{v}_1 = \begin{bmatrix} 1 \\ 0 \\ 0 \\ 0 \end{bmatrix}$, $\mathbf{v}_2 = \begin{bmatrix} 2 \\ 0 \\ 0 \\ 0 \end{bmatrix}$, $\mathbf{v}_3 = \begin{bmatrix} 0 \\ 1 \\ 0 \\ 0 \end{bmatrix}$, and $\mathbf{v}_4 = \begin{bmatrix} 5 \\ 3 \\ 0 \\ 0 \end{bmatrix}$. These are the columns of $RREF(A)$. Note that $\mathbf{v}_2 = 2\mathbf{v}_1$, and $\mathbf{v}_4 = 3\mathbf{v}_3 + 5\mathbf{v}_1$. Because elementary row operations do not change the solutions to equations, these same relations hold between the corresponding columns of the original matrix. We see that $\mathbf{v}_1$ and $\mathbf{v}_3$ are the columns of $RREF(A)$ which contain leading ones, $\mathbf{v}_1$ and $\mathbf{v}_3$ are obviously linearly independent, and the other columns $\mathbf{v}_2$ and $\mathbf{v}_4$ can be expressed as linear combinations of $\mathbf{v}_1$ and $\mathbf{v}_2$. These facts are also true of the corresponding columns of the original matrix A. Thus the columns of A corresponding to columns of $RREF(A)$ which have leading ones form a basis of the space spanned by the columns of A.

Procedure *To determine linear relations between the columns of a matrix, find the RREF of the matrix. The columns of the original matrix satisfy the same vector equations as the columns of the RREF.*

The subspace of R^4 spanned by the columns of A is called the **column space of** A. The subspace of R^4 spanned by the rows of A is called the **row space of** A.

Elementary row operations do not change the row space, since the new rows are linear combinations of the old rows and vice versa. Since the nonzero rows of $RREF(A)$ are obviously linearly independent (each has a 1 in a position where the others are all zero), the nonzero rows of $RREF(A)$ are a basis of the row space of A.

Procedure *To find a basis for the column space of a matrix A, find $RREF(A)$ and choose the columns of A corresponding to columns of $RREF(A)$ with leading ones.*

Procedure To find a basis for the row space of A, find $RREF(A)$. The nonzero rows of $RREF(A)$ are a basis of the row space of A.

Example Let
$$A = \begin{bmatrix} 2 & 4 & 3 & 19 \\ 3 & 6 & -1 & 12 \\ -1 & -2 & 2 & 1 \\ 4 & 8 & -6 & 2 \end{bmatrix}.$$

Then
$$RREF(A) = \begin{bmatrix} 1 & 2 & 0 & 5 \\ 0 & 0 & 1 & 3 \\ 0 & 0 & 0 & 0 \\ 0 & 0 & 0 & 0 \end{bmatrix}.$$

Thus a basis for the column space of A is

$$\begin{bmatrix} 2 \\ 3 \\ -1 \\ 4 \end{bmatrix} \text{ and } \begin{bmatrix} 3 \\ -1 \\ 2 \\ -6 \end{bmatrix}.$$

A basis for the row space of A is

$$(1, 2, 0, 5) \text{ and } (0, 0, 1, 3).$$

You can check that the columns of A, call them $\mathbf{u}_1$, $\mathbf{u}_2$, $\mathbf{u}_3$, and $\mathbf{u}_4$ satisfy the relations $\mathbf{u}_2 = 2\mathbf{u}_1$ and $\mathbf{u}_4 = 3\mathbf{u}_3 + 5\mathbf{u}_1$, as would the columns of any matrix which has this same RREF.

Remark Note that **elementary row operations often change the column space of a matrix**, so that it is important to go back to the original matrix A for a basis of the column space of A

Example Let
$$M = \begin{bmatrix} 1 & 2 & 1 \\ 3 & 6 & 3 \\ 0 & 0 & 0 \end{bmatrix}.$$

The row space of M is the set of all multiples of $(1, 2, 1)$, or equivalently all vectors of the form $(a, 2a, a)$. The column space of M is all multiples of $\begin{bmatrix} 1 \\ 3 \\ 0 \end{bmatrix}$, or equivalently all vectors of the form $(b, 3b, 0)$. (We notice that the column space and row space are different subspaces of R^3.) Now perform a row operation on M. Add -3 times the first row to the second row. The resulting matrix is

$$RREF(M) = \begin{bmatrix} 1 & 2 & 1 \\ 0 & 0 & 0 \\ 0 & 0 & 0 \end{bmatrix}.$$

The row space of $RREF(M)$ is still the same as the row space of M, but the column space of $RREF(M)$ is not the same as the column space of M, since the column space of $RREF(M)$ is all multiples of $\begin{bmatrix} 1 \\ 0 \\ 0 \end{bmatrix}$, or equivalently all vectors of the form $(c, 0, 0)$. A basis of the column space of M is $(1, 3, 0)$, and a basis of the row space of M is $(1, 2, 1)$.

Exercises

1. Find the RREF of the following matrices:

(a) $\begin{bmatrix} 1 & 2 \\ 2 & 1 \end{bmatrix}$

(b) $\begin{bmatrix} 1 & 1 & 1 \\ 0 & 1 & 2 \\ 0 & 1 & 3 \end{bmatrix}$

(c) $\begin{bmatrix} -3 & 1 & 1 & 1 \\ 1 & -3 & 1 & 1 \\ 1 & 1 & -3 & 1 \\ 1 & 1 & 1 & -3 \end{bmatrix}$

(d) $\begin{bmatrix} 1 & 2 & 6 \\ 3 & 6 & 1 \\ 4 & 8 & -1 \end{bmatrix}$

(e) $\begin{bmatrix} 4 & 1 & 1 & 1 \\ 1 & 4 & 1 & 1 \\ 1 & 1 & 4 & 1 \\ 1 & 1 & 1 & 4 \end{bmatrix}$

(f) $\begin{bmatrix} 2 & 2 & 4 & 12 \\ 2 & 1 & 5 & 8 \end{bmatrix}$

(g) $\begin{bmatrix} 3 & 2 \\ 1 & 5 \\ 0 & 6 \\ 1 & 1 \end{bmatrix}$.

2. For each of the matrices in problem 1, determine if the column vectors are linearly independent. Whenever they are linearly dependent, give a dependence relation (a dependence relation on a set of vectors is a vector equation showing the zero vector equal to a linear combination of the given vectors with not all the coefficients being zero.)

3. Using each matrix A of problem 1, determine whether $AX = \mathbf{O}$ has a nonzero solution. If it does, give such a solution.

4. Does $\begin{bmatrix} 1 & 2 & 3 \\ 2 & 3 & 4 \\ 3 & 4 & 5 \end{bmatrix} \begin{bmatrix} x \\ y \\ z \end{bmatrix} = \begin{bmatrix} 0 \\ 0 \\ 0 \end{bmatrix}$ have a nonzero solution? If the answer is yes, give such a solution.

5. Let $A = \begin{bmatrix} 1 & 2 & 1 & 2 \\ 2 & 4 & 2 & 4 \\ 0 & 0 & 0 & 0 \\ 3 & 6 & 5 & 8 \end{bmatrix}$. Find a basis for the column space of A and a basis for the row space of A.

6. Let $A = \begin{bmatrix} 1 & 3 & 1 & -2 \\ 1 & 4 & 3 & -1 \\ 2 & 3 & -4 & -7 \\ 3 & 8 & 1 & -7 \end{bmatrix}$. $\mathrm{RREF}(A) = \begin{bmatrix} 1 & 0 & -5 & -5 \\ 0 & 1 & 2 & 1 \\ 0 & 0 & 0 & 0 \\ 0 & 0 & 0 & 0 \end{bmatrix}$.

Find

(a) A basis of the row space of A.

(b) A basis of the column space of A.

(c) Express column 3 of A as a linear combination of columns 1 and 2 of A.

(d) Express column 4 of A as a linear combination of columns 1 and 2 of A.

7. Let $A = \begin{bmatrix} \mathbf{u}_1 & \mathbf{u}_2 & \mathbf{u}_3 & \mathbf{u}_4 \end{bmatrix}$, where $\mathbf{u}_1, \mathbf{u}_2, \mathbf{u}_3,$ and $\mathbf{u}_4$ are vectors in R^4 written as columns. If $RREF(A) = \begin{bmatrix} 1 & 0 & -2 & 0 \\ 0 & 1 & 3 & 0 \\ 0 & 0 & 0 & 1 \\ 0 & 0 & 0 & 0 \end{bmatrix}$, find:

(a) A basis of row space A.

(b) A basis of column space A.

(c) If possible, express $\mathbf{u}_3$ as a linear combination of $\mathbf{u}_1$ and $\mathbf{u}_2$.

(d) If possible express $\mathbf{u}_4$ as a linear combination of $\mathbf{u}_1$ and $\mathbf{u}_2$.

8. Determine if the following sets of vectors are linearly independent:

(a) $\{(1, 1, 2, -1), (3, -1, -2, 5), (5, 6, 12, 0), (0, 4, 8, 1)\}$

(b) $\{(1, 3, 2), (8, 10, 9), (0, 2, 1)\}$

(c) $\{(1, 2, 1), (0, 1, 2), (1, 3, 4)\}$

(d) $\{(1, 0, 0), (0, 1, 0), (0, 0, 1)\}$

(e) $(0, 1, 1), (1, 0, 1), (1, 1, 0)\}$.

Section 1.4 Inverse

An $n \times n$ (square) matrix A is invertible if there exists an nxn matrix B such that $AB = I = BA$. B is called the **inverse** of A. Some matrices have no inverse.

Example Let
$$A = \begin{bmatrix} 1 & 4 & -3 \\ 2 & 0 & 1 \\ 0 & 0 & 0 \end{bmatrix}.$$
Then for any 3×3 matrix B, the last row of AB will be all zeros, and thus AB cannot be I, so A has no inverse.

If a matrix A has two inverses, B and C, then we have
$$AB = I = BA$$
and also
$$AC = I = CA.$$
But then
$$B = IB = (CA)B = C(AB) = CI = C,$$
so that B must equal C. Thus a matrix A has at most one inverse. If A has an inverse we denote it by A^{-1}.

Fact For $n \times n$ matrices A and B, if $AB = I$ then $BA = I$ also. Thus you only need to check an inverse on one side.

Fact A square matrix A has an inverse if and only if $RREF(A) = I$.

Since I has a leading one in every column, $RREF(A) = I$ is equivalent to columns of A being linearly independent. Thus we also have:

Fact A square matrix A has an inverse if and only if the columns of A are linearly independent.

Procedure To find the inverse of an $n \times n$ matrix A, place the $n \times n$ identity matrix I next to A forming the $n \times (2n)$ matrix $\begin{bmatrix} A & I \end{bmatrix}$. Perform elementary row operations that change A into its $RREF$. If the A part becomes the identity matrix, the I part will become A^{-1}, so that $\begin{bmatrix} A & I \end{bmatrix}$ will change into $\begin{bmatrix} I & A^{-1} \end{bmatrix}$. If you get a zero row in the A part at any time, STOP! A is not invertible.

Example Invert $\begin{bmatrix} 2 & 3 \\ 1 & 2 \end{bmatrix}$. Form $\begin{bmatrix} 2 & 3 & 1 & 0 \\ 1 & 2 & 0 & 1 \end{bmatrix}$. The following sequence of elementary row operations reduces the first half of the matrix to I and the second half to the inverse of $\begin{bmatrix} 2 & 3 \\ 1 & 2 \end{bmatrix}$.

$$\begin{bmatrix} 2 & 3 & 1 & 0 \\ 1 & 2 & 0 & 1 \end{bmatrix} \to \begin{bmatrix} 1 & 2 & 0 & 1 \\ 2 & 3 & 1 & 0 \end{bmatrix} \to \begin{bmatrix} 1 & 2 & 0 & 1 \\ 0 & -1 & 1 & -2 \end{bmatrix}$$

$$\to \begin{bmatrix} 1 & 2 & 0 & 1 \\ 0 & 1 & -1 & 2 \end{bmatrix} \to \begin{bmatrix} 1 & 0 & 2 & -3 \\ 0 & 1 & -1 & 2 \end{bmatrix}.$$

Thus
$$\begin{bmatrix} 2 & 3 \\ 1 & 2 \end{bmatrix}^{-1} = \begin{bmatrix} 2 & -3 \\ -1 & 2 \end{bmatrix}.$$

Example Try to invert $\begin{bmatrix} 1 & 2 \\ 2 & 4 \end{bmatrix}$. Form $\begin{bmatrix} 1 & 2 & 1 & 0 \\ 2 & 4 & 0 & 1 \end{bmatrix}$. Adding -2 times the first row to the second row gives $\begin{bmatrix} 1 & 2 & 1 & 0 \\ 0 & 0 & -2 & 1 \end{bmatrix}$. STOP! $\begin{bmatrix} 1 & 2 \\ 2 & 4 \end{bmatrix}$ does not have an inverse since its RREF is $\begin{bmatrix} 1 & 2 \\ 0 & 0 \end{bmatrix}$.

Example Invert (find the inverse of) $\begin{bmatrix} 1 & 1 & 1 \\ 0 & 2 & 2 \\ 0 & 0 & 1 \end{bmatrix}$. Solution:

$$\begin{bmatrix} 1 & 1 & 1 & 1 & 0 & 0 \\ 0 & 2 & 2 & 0 & 1 & 0 \\ 0 & 0 & 1 & 0 & 0 & 1 \end{bmatrix} \to \begin{bmatrix} 1 & 1 & 1 & 1 & 0 & 0 \\ 0 & 1 & 1 & 0 & \frac{1}{2} & 0 \\ 0 & 0 & 1 & 0 & 0 & 1 \end{bmatrix} \to$$

$$\begin{bmatrix} 1 & 0 & 0 & 1 & -\frac{1}{2} & 0 \\ 0 & 1 & 1 & 0 & \frac{1}{2} & 0 \\ 0 & 0 & 1 & 0 & 0 & 1 \end{bmatrix} \to \begin{bmatrix} 1 & 0 & 0 & 1 & -\frac{1}{2} & 0 \\ 0 & 1 & 0 & 0 & \frac{1}{2} & -1 \\ 0 & 0 & 1 & 0 & 0 & 1 \end{bmatrix}.$$

Thus
$$\begin{bmatrix} 1 & 1 & 1 \\ 0 & 2 & 2 \\ 0 & 0 & 1 \end{bmatrix}^{-1} = \begin{bmatrix} 1 & -\frac{1}{2} & 0 \\ 0 & \frac{1}{2} & -1 \\ 0 & 0 & 1 \end{bmatrix}.$$

One application of inverting matrices is the solution of a system of n linear equations in n unknowns. This can be represented by the matrix equation $A\mathbf{x} = \mathbf{b}$, where $\mathbf{b}$ is a column vector and A is an $n \times n$ matrix. If A is invertible there is a unique solution and in fact the solution is $\mathbf{x} = A^{-1}\mathbf{b}$. Of course as we have seen you can also solve the equations by finding the RREF of the augmented matrix $\begin{bmatrix} A & \mathbf{b} \end{bmatrix}$. If A is not invertible this latter method is your only choice, and in this case there may be no solution or there may be many solutions.

In the case of 2×2 matrices, there is a formula for the inverse that you can use instead of the row reduction method illustrated above.

Formula For the Inverse of a 2×2 Matrix.
If
$$A = \begin{bmatrix} a & b \\ c & d \end{bmatrix}$$
and $ad - bc \neq 0$, then
$$A^{-1} = \frac{1}{ad-bc}\begin{bmatrix} d & -b \\ -c & a \end{bmatrix} = \begin{bmatrix} \frac{d}{ad-bc} & \frac{-b}{ad-bc} \\ \frac{-c}{ad-bc} & \frac{a}{ad-bc} \end{bmatrix}.$$

You can check by multiplying AA^{-1}. If $ad - bc = 0$, then A is not invertible.

Elementary Matrices
Any matrix obtained from the identity matrix I by one elementary row operation is called an **elementary matrix**.

Fact If E is an $n \times n$ elememtary matrix that comes from I by a certain elementary row operation, then for any $n \times k$ matrix A, EA is the matrix that comes from A by that same elementary row operation.

Example $\begin{bmatrix} 0 & 1 \\ 1 & 0 \end{bmatrix}\begin{bmatrix} 1 & 2 \\ 3 & 4 \end{bmatrix} = \begin{bmatrix} 3 & 4 \\ 1 & 2 \end{bmatrix}.$

Example $\begin{bmatrix} 1 & 0 \\ 0 & 3 \end{bmatrix}\begin{bmatrix} 1 & 2 \\ 3 & 4 \end{bmatrix} = \begin{bmatrix} 1 & 2 \\ 9 & 12 \end{bmatrix}.$

Vectors and Matrices

Example $\begin{bmatrix} 1 & 2 \\ 0 & 1 \end{bmatrix} \begin{bmatrix} 1 & 2 \\ 3 & 4 \end{bmatrix} = \begin{bmatrix} 7 & 10 \\ 3 & 4 \end{bmatrix}.$

You should be able to figure out which elementary row operation is involved in each of the three examples above.

Optional

We give a proof of the following fact:

Fact *A square matrix A is invertible if and only if $RREF(A) = I$.*

Proof. Since any elementary row operation on A can be performed by left multiplication of A by an elementary matrix, and since an arbitrary matrix A can be changed into a reduced row echelon form matrix by a sequence of elementary row operations, we can write

$$E_k E_{k-1} \cdots E_2 E_1 A = RREF(A),$$

where each E_i is an elementary matrix. If $RREF(A) = I$, this equation becomes

$$E_k E_{k-1} \cdots E_2 E_1 A = I.$$

An elementary matrix is invertible, and the inverse is an elementary matrix of the same type. (You can convince yourself of this by taking up each type separately.) Thus multiplying the previous equation on the left first by E_k^{-1}, then by E_{k-1}^{-1}, etc. leads to

$$A = E_1^{-1} E_2^{-1} \cdots E_{k-1}^{-1} E_k^{-1}.$$

The preceding equation shows that A is a product of elementary matrices. But then

$$A(E_k \cdots E_2 E_1) = (E_1^{-1} E_2^{-1} \cdots E_k^{-1})(E_k \cdots E_2 E_1) = I,$$

so A is invertible. Thus we have proved the fact that if $RREF(A) = I$, then A is invertible. If $RREF(A)$ is not I, then since A is square, $RREF(A)$ must have a zero row. Recall that if a matrix C has a zero row, then so does the product CD for every matrix D for which the multiplication is possible. For any square matrix B we have

$$E_k \cdots E_2 E_1 AB = RREF(A) \cdot B = M,$$

and since $RREF(A)$ has a zero row, M must have a zero row. If $AB = I$, this gives

$$E_k \cdots E_2 E_1 I = M,$$

so that

$$I = M(E_1^{-1} E_2^{-1} \cdots E_k^{-1}),$$

a contradiction because the matrix on the right hand side of this equation has a zero row. Thus we have shown that if A is invertible and $RREF(A) \neq I$, we get a contradiction. Thus if A is invertible we must have $RREF(A) = I$. ∎

Along the way in the above argument we also showed that an invertible matrix is equal to a product of elementary matrices.

It can also be seen from the equation

$$E_k \cdots E_2 E_1 A = I$$

why our method of computing inverses works. If we multiply both sides of this equation by A^{-1} on the right, we get the equation

$$E \cdots E_2 E_1 I = A^{-1},$$

which says that the same sequence of row operations that change A into I will change I into A^{-1}.

Exercises

1. Determine whether each of the following matrices is invertible and if it is, find its inverse.

    (a) $\begin{bmatrix} 3 & 2 \\ 5 & 3 \end{bmatrix}$

    (b) $\begin{bmatrix} 1 & p \\ p & 1 \end{bmatrix}$

Vectors and Matrices

(c) $\begin{bmatrix} 1 & 2 & 3 \\ 2 & 3 & 4 \\ 3 & 4 & 5 \end{bmatrix}$

(d) $\begin{bmatrix} 0 & 1 & 2 \\ 0 & 0 & 3 \\ 0 & 0 & 0 \end{bmatrix}$

(e) $\begin{bmatrix} 1 & 2 & 4 \\ 1 & 3 & 9 \\ 1 & 4 & 16 \end{bmatrix}$

(f) $\begin{bmatrix} 4 & 1 & 1 \\ 1 & 4 & 1 \\ 1 & 1 & 4 \end{bmatrix}$.

2. A matrix A is said to be **idempotent** if $A^2 = A$. Determine all 3×3 idempotent matrices that are invertible. Can you generalize this to $n \times n$ idempotent matrices?

3. Show that if A and B are invertible $n \times n$ matrices, then so are AB and BA.

4. Show by example that each type of elementary matrix has an inverse.

5. Using matrix inversion, solve the following system of linear equations:
$$\begin{aligned} x + 2y + 4z &= 8 \\ x + 3y + 9z &= 14 \\ x + 4y + 16z &= 22 \end{aligned}.$$

Section 1.5 Rank and Dimension

A very important fact about subspaces of R^n is the following:

Fact *If a subspace W of R^n has a basis with k vectors, then every basis of W contains exactly k vectors.*

The **dimension** of a subspace W is the number of vectors in a basis of W. This number will be the same no matter which basis of W we use.

The row vectors of the $n \times n$ identity matrix, $(1, 0, \cdots, 0)$, $(0, 1, 0, \cdots, 0)$ $\cdots$, $(0, \cdots, 0, 1)$, are a basis of R^n, called the **standard basis of $\mathbf{R}^n$**, thus the dimension of R^n is n, and every basis of R^n has exactly n vectors.

Recall from Section 1.3 that we can find a basis of the column space of a matrix A by choosing the columns of A corresponding to the columns of $RREF(A)$ which contain leading ones. Therefore the dimension of the column space of A equals the number of leading ones in $RREF(A)$. We also saw that the nonzero rows of $RREF(A)$ are a basis for the row space of A. Therefore the dimension of the row space of A also equals the number of leading ones in $RREF(A)$. This gives the important result:

Fact *For every matrix A, dimension row space = dimension column space.*

The **rank** of a matrix A is the number of leading ones in $RREF(A)$. From the above we see that $rank(A) =$ dimension row space $A =$ dimension column space A.

For an $n \times n$ matrix A we can see from the above fact that the rows of A are linearly independent if and only if the columns of A are linearly independent. We already know the columns are linearly independent if and only if $RREF(A) = I$. It is clear from the definition of rank that an $n \times n$ matrix A has rank n if and only if $RREF(A) = I$. We can also connect the concept of rank with invertibility.

Fact *An $n \times n$ matrix A is invertible if and only if $rank(A) = n$.*

Reason. Both statements about A are equivalent to $RREF(A) = I$. ∎

Example Let
$$A = \begin{bmatrix} 1 & -1 & 0 & -3 \\ 2 & 4 & 2 & 12 \\ 0 & 2 & 0 & 6 \\ 0 & 3 & 5 & 9 \end{bmatrix}.$$

Then
$$RREF(A) = \begin{bmatrix} 1 & 0 & 0 & 0 \\ 0 & 1 & 0 & 3 \\ 0 & 0 & 1 & 0 \\ 0 & 0 & 0 & 0 \end{bmatrix}.$$

$Rank(A) = 3$.

Procedure *To find the dimension of a subspace W of R^n given a spanning set of W, put the spanning set vectors as either rows or columns of a matrix and find the rank of that matrix. This is the dimension of W.*

Example *Let*

$$W = sp\{(3,2,5,3), (0,1,1,0), (3,4,7,3)\},$$

where $sp\{\mathbf{u}_1, \mathbf{u}_2, \cdots, \mathbf{u}_k\}$ stands for the set of all linear combinations of $\mathbf{u}_1, \mathbf{u}_2, \cdots, \mathbf{u}_k$, i.e. span of $\mathbf{u}_1, \mathbf{u}_2, \cdots, \mathbf{u}_k$. Find a basis of W and the dimension of W. Solution: We can put the vectors spanning W as either rows or columns of a matrix. Let

$$A = \begin{bmatrix} 3 & 0 & 3 \\ 2 & 1 & 4 \\ 5 & 1 & 7 \\ 3 & 0 & 3 \end{bmatrix},$$

then

$$RREF(A) = \begin{bmatrix} 1 & 0 & 1 \\ 0 & 1 & 2 \\ 0 & 0 & 0 \\ 0 & 0 & 0 \end{bmatrix}.$$

Thus we see that the dimension of W is 2, and a basis of W is

$$\{(3,2,5,3), (0,1,1,0)\}.$$

If we had put the vectors as rows,

$$B = \begin{bmatrix} 3 & 2 & 5 & 3 \\ 0 & 1 & 1 & 0 \\ 3 & 4 & 7 & 3 \end{bmatrix},$$

then
$$RREF(B) = \begin{bmatrix} 1 & 0 & 1 & 1 \\ 0 & 1 & 1 & 0 \\ 0 & 0 & 0 & 0 \end{bmatrix}.$$

Again we get 2 as the answer for the dimension, but we get a different basis for W, namely a basis of W is

$$\{(1,0,1,1),(0,1,1,0)\}.$$

If you want to select a basis from the vectors of the original spanning set, use the first method, putting the spanning set as columns.

Fact *Any spanning set of a subspace W of R^n contains a subset which is a basis of W.*

Procedure *Put the spanning set as columns of a matrix A. The columns of A corresponding to columns of $RREF(A)$ containing leading ones are a basis of W.*

Fact *Any linearly independent set in a subspace W of R^n can be extended to a basis of W.*

Procedure *Put the linearly independent set of vectors from W as the first columns of a matrix and then put a spanning set of W as the rest of the columns. Select a basis of the column space in the usual way. This basis will include the given linearly independent set.*

Example *Extend $\{(2,0,1,1),(1,1,1,1)\}$ to a basis of R^4. This set is linearly independent since $(1,1,1,1) \neq c(2,0,1,1)$ for every c. We can use as our spanning set of R^4 the standard basis. Form the matrix*

$$A = \begin{bmatrix} 2 & 1 & 1 & 0 & 0 & 0 \\ 0 & 1 & 0 & 1 & 0 & 0 \\ 1 & 1 & 0 & 0 & 1 & 0 \\ 1 & 1 & 0 & 0 & 0 & 1 \end{bmatrix}.$$

$$RREF(A) = \begin{bmatrix} 1 & 0 & 0 & -1 & 0 & 1 \\ 0 & 1 & 0 & 1 & 0 & 0 \\ 0 & 0 & 1 & 1 & 0 & -2 \\ 0 & 0 & 0 & 0 & 1 & -1 \end{bmatrix}.$$

Vectors and Matrices

The leading ones in $RREF(A)$ occur in columns 1,2,3, and 5, Thus columns 1,2,3, nd 5 of A are a basis of R^4. Thus the vectors $(2,0,1,1)$, $(1,1,1,1)$, $(1,0,0,0)$, $(0,0,1,0)$ form a basis of R^4 that includes $(2,0,1,1)$ and $(1,1,1,1)$.

Remark *In the next section we will have another (better) way to do the above problem.*

Fact *If the dimension of a subspace W is k, then*

1. *the number of elements in a spanning set of W is $\geq k$*

   the number of elements in a linearly independent set in W is $\leq k$

   the number of elements in a basis of W is k.

   k vectors in W are a basis of W if and only if they are linearly independent.

Fact *If $\{\mathbf{v}_1, \mathbf{v}_2, \cdots \mathbf{v}_k\}$ is a basis of W, then every vector in W can be expressed uniquely as a linear combination of $\mathbf{v}_1, \mathbf{v}_2, \cdots \mathbf{v}_k$.*

Reason. Recall that a basis of W is a spanning set of W that is also a linearly independent set. Because it is a spanning set, every vector in W is a linear combination of $\mathbf{v}_1, \mathbf{v}_2, \cdots \mathbf{v}_k$. The uniqueness comes from the linear independence. If we had two ways to write a vector w, say

$$w = a_1\mathbf{v}_1 + a_2\mathbf{v}_2 + \cdots + a_k\mathbf{v}_k,$$

and

$$w = b_1\mathbf{v}_1 + b_2\mathbf{v}_2 + \cdots + b_k\mathbf{v}_k,$$

then subtracting gives

$$\mathbf{0} = (a_1 - b_1)\mathbf{v}_1 + (a_2 - b_2)\mathbf{v}_2 + \cdots (a_k - b_k)\mathbf{v}_k,$$

and since $\mathbf{v}_1, \mathbf{v}_2, \cdots \mathbf{v}_k$ are linearly independent all these coefficients must be 0, so that $a_i = b_i$ for all $i = 1, 2, \cdots, k$.

Exercises

1. Find the rank of the following matrices:

    (a) $\begin{bmatrix} 2 & 1 \\ 1 & 2 \end{bmatrix}$

    (b) $\begin{bmatrix} 3 & 1 & 1 \\ 1 & 3 & 1 \\ 1 & 1 & 3 \end{bmatrix}$

    (c) $\begin{bmatrix} -3 & 1 & 1 & 1 \\ 1 & -3 & 1 & 1 \\ 1 & 1 & -3 & 1 \\ 1 & 1 & 1 & -3 \end{bmatrix}$

    (d) $\begin{bmatrix} 1 & 1 & 1 \\ 1 & 2 & 3 \\ 3 & 1 & 0 \end{bmatrix}$

    (e) $\begin{bmatrix} 4 & 1 & 1 & 1 \\ 1 & 4 & 1 & 1 \\ 1 & 1 & 4 & 1 \\ 1 & 1 & 1 & 4 \end{bmatrix}.$

2. Extend the following linearly independent sets of vectors to a basis of R^n for the appropriate n.

    (a) $\{(2,1,1),(1,2,1)\}$
    (b) $\{(1,0,1,0),(1,1,1,1)\}$
    (c) $\{(1,2,3,4),(4,3,2,1)\}$.

3.

   (a) Express $(1,2,3)$ as a linear combination of the basis you obtained in 2a.

   (b) Express $(1,2,3,4)$ as a linear combination of the basis you obtained in 2b.

   (c) Express $(1,0,0,0)$ as a linear combination of the basis you obtained in 2c.

4. Determine all values of q for which the matrix $\begin{bmatrix} q & 1 & 1 \\ 2 & 1 & 3 \\ 0 & 5 & 1 \end{bmatrix}$ is invertible.

Section 1.6 Null Space and Hermite Form

Fact *Given any $m \times n$ matrix A, the set of all $n \times 1$ matrices (column vectors) $\mathbf{x}$ such that $A\mathbf{x} = \mathbf{O}$ forms a subspace of R^n.*

Reason. If $A\mathbf{x}_1 = \mathbf{O}$ and $A\mathbf{x}_2 = \mathbf{O}$, then $A(\mathbf{x}_1 + \mathbf{x}_2) = A\mathbf{x}_1 + A\mathbf{x}_2 = \mathbf{O} + \mathbf{O} = \mathbf{O}$. Also for any scalar c, $A(c\mathbf{x}) = c(A\mathbf{x}) = c\mathbf{O} = \mathbf{O}$.

The **null space of A** is the set of all vectors $\mathbf{x}$ satisfying $A\mathbf{x} = \mathbf{O}$. The **nullity of A** is the dimension of the null space of A.

We need to be able to find a basis for the null space of A, and the following will lead to a relatively easy way to do this.

Hermite Form of A

When A is $m \times n$, $RREF(A)$ cannot have more than n leading ones. If there are more rows, they must be zero rows. If $m > n$, delete $m - n$ zero rows from $RREF(A)$. If $m < n$, add $n - m$ zero rows to $RREF(A)$. In either case call the resulting $n \times n$ matrix C. Now rearrange the rows of C so that the leading ones all occur on the main (left to right) diagonal. The resulting matrix is the **Hermite form of A** which we denote by $H(A)$. It is uniquely determined by A.

Example *Suppose*

$$RREF(A) = \begin{bmatrix} 1 & 2 & 0 & 0 & 5 \\ 0 & 0 & 1 & 0 & 3 \\ 0 & 0 & 0 & 1 & 6 \end{bmatrix}.$$

Then
$$C = \begin{bmatrix} 1 & 2 & 0 & 0 & 5 \\ 0 & 0 & 1 & 0 & 3 \\ 0 & 0 & 0 & 1 & 6 \\ 0 & 0 & 0 & 0 & 0 \\ 0 & 0 & 0 & 0 & 0 \end{bmatrix},$$

and

$$H(A) = \begin{bmatrix} 1 & 2 & 0 & 0 & 5 \\ 0 & 0 & 0 & 0 & 0 \\ 0 & 0 & 1 & 0 & 3 \\ 0 & 0 & 0 & 1 & 6 \\ 0 & 0 & 0 & 0 & 0 \end{bmatrix}.$$

Recall that the matrix Q is idempotent if $Q^2 = Q$. It can be shown that for every matrix A, the Hermite form $H(A)$ is an idempotent matrix. Also A and $H(A)$ have the same null space, since neither adding zero rows nor elementary row operations change the solutions of the system represented by $A\mathbf{x} = \mathbf{O}$. Let $H = H(A)$. Then $H(H-I) = H^2 - H = [0]$. Thus all the column vectors of $H - I$ are in the null space of H and therefore also in the null space of A.

Fact *The nonzero column vectors of $H - I$ form a basis of the null space of A.*

If $rank(A) = k$, then $RREF(A)$ has k leading ones. Then H also has k leading ones, and these are on the main diagonal, so when we subtract I, the columns where H had a leading one become zero columns. Thus the number of nonzero columns in $H - I$ is $n - k$.

Fact $Rank(A) + Nullity(A) = $ *number of columns of A.*

Example Let $A = \begin{bmatrix} 1 & 2 & 4 & 3 \\ -2 & 0 & -4 & 1 \\ 1 & 3 & 5 & 0 \\ 1 & 1 & 3 & 1 \end{bmatrix}$. Find a basis of the null space of A. Solution:

$$RREF(A) = \begin{bmatrix} 1 & 0 & 2 & 0 \\ 0 & 1 & 1 & 0 \\ 0 & 0 & 0 & 1 \\ 0 & 0 & 0 & 0 \end{bmatrix},$$

Vectors and Matrices

so
$$H = \begin{bmatrix} 1 & 0 & 2 & 0 \\ 0 & 1 & 1 & 0 \\ 0 & 0 & 0 & 0 \\ 0 & 0 & 0 & 1 \end{bmatrix},$$

and
$$H - I = \begin{bmatrix} 0 & 0 & 2 & 0 \\ 0 & 0 & 1 & 0 \\ 0 & 0 & -1 & 0 \\ 0 & 0 & 0 & 0 \end{bmatrix}.$$

A basis of the null space is $\{(2, 1, -1, 0)\}$. $Rank(A) = 3$, $Nullity(A) = 1$.

Example Let
$$A = \begin{bmatrix} 1 & 3 & 2 & 0 \\ 0 & 2 & 2 & 1 \end{bmatrix}.$$

Then
$$RREF(A) = \begin{bmatrix} 1 & 0 & -1 & -\frac{3}{2} \\ 0 & 1 & 1 & \frac{1}{2} \end{bmatrix}.$$

So
$$H = \begin{bmatrix} 1 & 0 & -1 & -\frac{3}{2} \\ 0 & 1 & 1 & \frac{1}{2} \\ 0 & 0 & 0 & 0 \\ 0 & 0 & 0 & 0 \end{bmatrix},$$

and
$$H - I = \begin{bmatrix} 0 & 0 & -1 & -\frac{3}{2} \\ 0 & 0 & 1 & \frac{1}{2} \\ 0 & 0 & -1 & 0 \\ 0 & 0 & 0 & -1 \end{bmatrix}.$$

Thus a basis for the null space of A is

$$\{(-1, 1, -1, 0), (-\frac{3}{2}, \frac{1}{2}, 0, -1)\}.$$

$Rank(A) = 2$, $Nullity(A) = 2$.

A new method for extending a linearly independent set of vectors in R^n to a basis of R^n is based on the following fact.

Fact *A basis of the row space of A together with a basis for the null space of A forms a basis for R^n (where n is the number of columns of A).*

Example *Find a basis of R^3 containing the vector $(0,1,4)$. Solution:* Let
$$A = \begin{bmatrix} 0 & 1 & 4 \end{bmatrix}.$$
Then
$$RREF(A) = \begin{bmatrix} 0 & 1 & 4 \end{bmatrix},$$
and
$$H = \begin{bmatrix} 0 & 0 & 0 \\ 0 & 1 & 4 \\ 0 & 0 & 0 \end{bmatrix}.$$

$$H - I = \begin{bmatrix} -1 & 0 & 0 \\ 0 & 0 & 4 \\ 0 & 0 & -1 \end{bmatrix}.$$

Thus a basis for the null space of A is
$$\{(-1,0,0),(0,4,-1)\},$$
and
$$\{(0,1,4),(-1,0,0),(0,4,-1)\}$$
is a basis of R^3.

Example *Extend the set*
$$\{(1,2,-1,2,14),(2,4,1,-2,1),(3,6,5,-5,0)\}$$
to a basis of R^5. Solution: Form
$$A = \begin{bmatrix} 1 & 2 & -1 & 2 & 14 \\ 2 & 4 & 1 & -2 & 1 \\ 2 & 6 & 5 & -5 & 0 \end{bmatrix}.$$

$$RREF(A) = \begin{bmatrix} 1 & 2 & 0 & 0 & 5 \\ 0 & 0 & 1 & 0 & 3 \\ 0 & 0 & 0 & 1 & 6 \end{bmatrix},$$

Vectors and Matrices

$$H = \begin{bmatrix} 1 & 2 & 0 & 0 & 5 \\ 0 & 0 & 0 & 0 & 0 \\ 0 & 0 & 1 & 0 & 3 \\ 0 & 0 & 0 & 1 & 6 \\ 0 & 0 & 0 & 0 & 0 \end{bmatrix},$$

and

$$H - I = \begin{bmatrix} 0 & 2 & 0 & 0 & 5 \\ 0 & -1 & 0 & 0 & 0 \\ 0 & 0 & 0 & 0 & 3 \\ 0 & 0 & 0 & 0 & 6 \\ 0 & 0 & 0 & 0 & -1 \end{bmatrix}.$$

The original set and $(2, -1, 0, 0, 0)$ and $(5, 0, 3, 6, -1)$ form a basis of R^5.

Procedure To extend a linearly independent set of vectors in R^n to a basis of R^n, put the vectors of the linearly independent set as the rows of a matrix A. Find a basis of the null space of A. The union of the two sets of vectors is a basis of R^n.

Exercises

1. Find a basis of the null space of the following matrices.

(a) $\begin{bmatrix} 1 & 2 & 3 \\ 4 & 5 & 6 \\ 7 & 8 & 9 \end{bmatrix}$

(b) $\begin{bmatrix} -2 & 1 & 1 \\ 1 & -2 & 1 \\ 1 & 1 & -2 \end{bmatrix}$

(c) $\begin{bmatrix} 1 & 2 & 1 \\ 0 & 1 & 2 \end{bmatrix}$

(d) $\begin{bmatrix} 1 & 1 & -2 & 0 \\ 0 & 1 & 2 & 3 \\ 0 & 0 & 0 & 1 \end{bmatrix}$

(e) $\begin{bmatrix} -3 & 1 & 1 & 1 \\ 1 & -3 & 1 & 1 \\ 1 & 1 & -3 & 1 \\ 1 & 1 & 1 & -3 \end{bmatrix}.$

2. For each of the matrices in problem 1, find a basis of the row space and then extend it to a basis of R^n (where n is the number of columns in the matrix).

3. If a matrix A is square and invertible what is the null space of A? Explain.

Chapter 2

Determinants

Section 2.1 Definition and Properties

When A is a square matrix there is a number associated with A called the **determinant** of A. If A is a 1×1 matrix $A = \begin{bmatrix} a \end{bmatrix}$, the determinant of A is the number a. One can define determinants for larger matrices inductively, in other words give an equation that expresses the determinant of an $n \times n$ matrix in terms of determinants of smaller matrices. If A is an $n \times n$ matrix, by a **minor** of A, denoted by M_{ij}, we mean the $(n-1) \times (n-1)$ matrix obtained from A by deleting row i and column j. We denote the determinant of A by either $det(A)$ or $|A|$. If we denote by a_{ij} the entry of A in row i, column j, then

$$|A| = a_{11}|M_{11}| - a_{12}|M_{12}| + a_{13}|M_{13}| - \cdots (-1)^n a_{1n}|M_{1n}|.$$

Remark *We alternate + and − in this formula.*

If
$$A = \begin{bmatrix} a_{11} & a_{12} \\ a_{21} & a_{22} \end{bmatrix},$$

then
$$\det(A) = \begin{vmatrix} a_{11} & a_{12} \\ a_{21} & a_{22} \end{vmatrix} = a_{11}a_{22} - a_{12}a_{21},$$
using the formula above for $n = 2$. When $n = 3$ we have

$$\begin{vmatrix} a_{11} & a_{12} & a_{13} \\ a_{21} & a_{22} & a_{23} \\ a_{31} & a_{32} & a_{33} \end{vmatrix} = a_{11}(a_{22}a_{33} - a_{23}a_{32})$$
$$-a_{12}(a_{21}a_{33} - a_{23}a_{31})$$
$$+a_{13}(a_{21}a_{32} - a_{22}a_{31}).$$

The number of terms in the formula for the determinant of an $n \times n$ matrix is n factorial, $n! = n(n-1)(n-2)\cdots(2)(1)$. This grows rapidly with n, so it is fortunate that there are properties of the determinant which allow us to compute it with fewer operations.

A matrix is called upper triangular if all entries below the main diagonal are zero. For example the matrix

$$\begin{bmatrix} 1 & 2 & 3 & 4 \\ 0 & 5 & 6 & 7 \\ 0 & 0 & 8 & 9 \\ 0 & 0 & 0 & 10 \end{bmatrix}$$

is upper triangular.

Fact *The determinant of an upper triangular matrix is the product of all elements on the main diagonal.*

Example *The determinant of the upper triangular matrix given above is $(1)(5)(8)(10) = 400$. Also for the identity matrix of any size $|I| = 1$, and for the zero matrix of any size, $|0| = 0$.*

The **transpose** of a matrix A, written A^T, is the matrix whose columns are the rows of A. The entry in the ij position of A^T is the entry in the ji position of A. Transpose is defined even if A is not square.

Determinants

Example *If*
$$A = \begin{bmatrix} 1 & 2 & 3 \\ 4 & 5 & 6 \\ 7 & 8 & 9 \end{bmatrix},$$

then
$$A^T = \begin{bmatrix} 1 & 4 & 7 \\ 2 & 5 & 8 \\ 3 & 6 & 9 \end{bmatrix}.$$

Example *If*
$$B = \begin{bmatrix} 1 & 2 \\ 4 & 5 \\ 7 & 8 \end{bmatrix},$$

then
$$B^T = \begin{bmatrix} 1 & 4 & 7 \\ 2 & 5 & 8 \end{bmatrix}.$$

Example *If*
$$C = \begin{bmatrix} 1 & 2 & 3 & 4 \\ 0 & 5 & 6 & 7 \\ 0 & 0 & 8 & 9 \\ 0 & 0 & 0 & 10 \end{bmatrix},$$

then
$$C^T = \begin{bmatrix} 1 & 0 & 0 & 0 \\ 2 & 5 & 0 & 0 \\ 3 & 6 & 8 & 0 \\ 4 & 7 & 9 & 10 \end{bmatrix}.$$

Fact *For all square matrices A, $\left|A^T\right| = |A|$.*

If E is an elementary matrix then $|EA|$ and $|A|$ are related, depending on which type of elementary row operation E performs.

Effect of Elementary Row Operations on Determinant

1. If E adds a multiple of one row to another then $|EA| = |A|$.

2. If E multiplies a row by $c \neq 0$, then $|EA| = c|A|$.

3. If E switches two rows, then $|EA| = -|A|$.

Example *Start with*

$$A = \begin{bmatrix} 1 & 2 & 3 \\ 4 & 5 & 6 \\ 7 & 8 & 9 \end{bmatrix}.$$

If E adds twice row one to row two, then

$$|EA| = \begin{vmatrix} 1 & 2 & 3 \\ 6 & 9 & 12 \\ 7 & 8 & 9 \end{vmatrix} = \begin{vmatrix} 1 & 2 & 3 \\ 4 & 5 & 6 \\ 7 & 8 & 9 \end{vmatrix}.$$

If E multiplies row one by two then

$$|EA| = \begin{vmatrix} 2 & 4 & 6 \\ 4 & 5 & 6 \\ 7 & 8 & 9 \end{vmatrix} = 2 \cdot \begin{vmatrix} 1 & 2 & 3 \\ 4 & 5 & 6 \\ 7 & 8 & 9 \end{vmatrix}.$$

If E switches row two and row three then

$$|EA| = \begin{vmatrix} 1 & 2 & 3 \\ 7 & 8 & 9 \\ 4 & 5 & 6 \end{vmatrix} = - \begin{vmatrix} 1 & 2 & 3 \\ 4 & 5 & 6 \\ 7 & 8 & 9 \end{vmatrix}.$$

This leads to a method for finding the determinant of a matrix.

Procedure *Perform elementary row operations on A until you get a matrix that is upper triangular. Keep track of all the changes that your row operations make on the determinant. Since you can easily find the determinant of the upper triangular matrix, you can now compute what the determinant of the original matrix must be.*

Determinants

Remark We know that we can always reach an upper triangular form, since $RREF(A)$ is upper triangular.

Remark Since $|A| = |A^T|$, we could use elementary column operations as well as elementary row operations when computing determinants. (This is something that you must not do when finding $RREF(A)$.)

If a row or column of A is the zero vector then $|A| = 0$. It is also clear that while elementary row operations may change $|A|$, they do not change it from zero to nonzero or vice versa. Thus $|A| = 0$ if and only if $RREF(A)$ has a zero row and $|A| \neq 0$ if and only if $RREF(A) = I$. Thus knowing if $|A|$ is zero or nonzero gives a lot of information.

Fact If $|A| = 0$, then the rows of A are linearly dependent, the columns of A are linearly dependent, A is not invertible, and the null space of A has dimension ≥ 1.

Fact If $|A| \neq 0$, then the rows of A are linearly independent, the columns of A are linearly independent, A is invertible, the rank of A is n, and the null space of A contains only the zero vector.

Another important property of determinants is the following

Fact $|AB| = |A| |B|$.

Using the definition of determinant and the fact that switching rows multiplies the determinant by -1, one can show that $|A|$ can be obtained from any row and its minors. Using the fact that $|A| = |A^T|$, it can be seen that $|A|$ can also be obtained from any column and its minors. The formulas are

Expansion by Minors

AROUND ROW i

$$|A| = (-1)^{i+1} a_{i1} |M_{i1}| + (-1)^{i+2} a_{i2} |M_{i2}| + \cdots + (-1)^{i+n} a_{in} |M_{in}|.$$

AROUND COLUMN j

$$|A| = (-1)^{1+j} a_{1j} |M_{1j}| + (-1)^{2+j} a_{2j} |M_{2j}| + \cdots + (-1)^{n+j} a_{nj} |M_{nj}|.$$

These formulas are especially useful once one obtains a matrix with a row or column in which only one entry is not zero.

Example

$$\begin{vmatrix} 1 & 2 & 3 & 4 \\ 0 & 0 & 1 & 0 \\ 5 & 6 & 7 & 8 \\ 9 & 8 & 7 & 6 \end{vmatrix} = (-1)^{2+3} \begin{vmatrix} 1 & 2 & 4 \\ 5 & 6 & 8 \\ 9 & 8 & 6 \end{vmatrix}$$
$$= (-1)(-28 - 2(-42) + 4(-14)) = 0.$$

Example

$$\begin{vmatrix} 1 & 0 & 5 & 9 \\ 2 & 0 & 6 & 8 \\ 3 & 1 & 7 & 7 \\ 4 & 0 & 8 & 6 \end{vmatrix} = (-1)^{3+2} \begin{vmatrix} 1 & 5 & 9 \\ 2 & 6 & 8 \\ 4 & 8 & 6 \end{vmatrix}$$
$$= (-1)(-28 - 5(-20) + 9 \cdot (-8)) = 0.$$

One can also use elementary row operations to find determinants.

$$\begin{vmatrix} 2 & 1 & 1 & 1 \\ 1 & 2 & 1 & 1 \\ 1 & 1 & 2 & 1 \\ 1 & 1 & 1 & 2 \end{vmatrix} = \begin{vmatrix} 3 & 3 & 2 & 2 \\ 1 & 2 & 1 & 1 \\ 1 & 1 & 2 & 1 \\ 1 & 1 & 1 & 2 \end{vmatrix} = \begin{vmatrix} 4 & 4 & 4 & 3 \\ 1 & 2 & 1 & 1 \\ 1 & 1 & 2 & 1 \\ 1 & 1 & 1 & 2 \end{vmatrix}$$
$$= \begin{vmatrix} 5 & 5 & 5 & 5 \\ 1 & 2 & 1 & 1 \\ 1 & 1 & 2 & 1 \\ 1 & 1 & 1 & 2 \end{vmatrix} = 5 \cdot \begin{vmatrix} 1 & 1 & 1 & 1 \\ 1 & 2 & 1 & 1 \\ 1 & 1 & 2 & 1 \\ 1 & 1 & 1 & 2 \end{vmatrix}$$
$$= 5 \cdot \begin{vmatrix} 1 & 1 & 1 & 1 \\ 0 & 1 & 0 & 0 \\ 1 & 1 & 2 & 1 \\ 1 & 1 & 1 & 2 \end{vmatrix} = 5 \cdot \begin{vmatrix} 1 & 1 & 1 & 1 \\ 0 & 1 & 0 & 0 \\ 0 & 0 & 1 & 0 \\ 1 & 1 & 1 & 2 \end{vmatrix}$$

Determinants

$$= 5 \cdot \begin{vmatrix} 1 & 1 & 1 & 1 \\ 0 & 1 & 0 & 0 \\ 0 & 0 & 1 & 0 \\ 0 & 0 & 0 & 1 \end{vmatrix} = 5 \cdot 1 \cdot 1 \cdot 1 \cdot 1 = 5.$$

Example

$$\begin{vmatrix} 1 & 2 & 3 & 4 \\ 2 & 3 & 4 & 5 \\ 3 & 4 & 5 & 6 \\ 4 & 5 & 6 & 7 \end{vmatrix} = \begin{vmatrix} 1 & 2 & 3 & 4 \\ 2 & 3 & 4 & 5 \\ 3 & 4 & 5 & 6 \\ 1 & 1 & 1 & 1 \end{vmatrix}$$

$$= \begin{vmatrix} 1 & 2 & 3 & 4 \\ 2 & 3 & 4 & 5 \\ 1 & 1 & 1 & 1 \\ 1 & 1 & 1 & 1 \end{vmatrix}$$

$$= \begin{vmatrix} 1 & 2 & 3 & 4 \\ 2 & 3 & 4 & 5 \\ 1 & 1 & 1 & 1 \\ 0 & 0 & 0 & 0 \end{vmatrix} = 0.$$

In the preceding example we used only elementary row operations that did not change the determinant, namely adding a multiple of one row to another. We could have saved one step by observing that in general

Fact *If a matrix has two identical rows (or two identical columns) then its determinant is zero.*

Exercises

1. Evaluate the following determinants:

(a) $\begin{vmatrix} 1 & 3 & 4 \\ 1 & 7 & -1 \\ 1 & 2 & 8 \end{vmatrix}$

(b) $\begin{vmatrix} 2 & 1 & 1 & 0 \\ 1 & 3 & 1 & 0 \\ 1 & 0 & 4 & 1 \\ 1 & 1 & 1 & 5 \end{vmatrix}$

(c) $\begin{vmatrix} -3 & 1 & 1 & 1 \\ 1 & -3 & 1 & 1 \\ 1 & 1 & -3 & 1 \\ 1 & 1 & 1 & -3 \end{vmatrix}$

(d) $\begin{vmatrix} x & y & x & x \\ x & y & y & y \\ y & y & x & y \\ x & x & x & y \end{vmatrix}$

(e) $\begin{vmatrix} 2-x & 1 & 0 & 0 \\ 1 & 2-x & 0 & 0 \\ 0 & 0 & 2-x & 1 \\ 0 & 0 & 2 & 3-x \end{vmatrix}$.

2. If A is a 4×4 matrix with $|A| = 11$, answer the following questions:

   (a) What is $RREF(A)$?
   (b) What is $rank(A)$?
   (c) What is the null space of A?
   (d) Is A invertible?
   (e) Are the rows of A a basis of R^4?
   (f) Find $|2A|$.

3. For each of the following set of vectors in R^3, use a determinant to decide whether or not the set is a basis of R^3.

   (a) $\{(1,1,1),(1,1,-2),(2,1,1)\}$
   (b) $\{(1,1,2),(2,3,1),(1,2,1)\}$
   (c) $\{(1,1,2),(2,3,1),(1,2,-1)\}$.

4. For what values of k are the vectors $(1,1,3)$, $(2,1,1)$, $(3,-1,k)$ linearly independent?

5. Use determinants to decide whether the following system of equations has a nontrivial (nonzero) solution.

$$\begin{aligned} x - 2y + z &= 0 \\ 2x + 3y + z &= 0 \\ 3x + y + 2z &= 0 \end{aligned}.$$

Determinants

6. Evaluate the determinant $\begin{vmatrix} e^x & 2e^x & 4e^x \\ e^{2x} & e^{2x} & e^{2x} \\ e^{-x} & -e^{-x} & e^{-x} \end{vmatrix}$.

7. Using the fact that $|AB| = |A||B|$, show that if A is invertible then
$$|A^{-1}| = \frac{1}{|A|}.$$

8. Let A and B be square matrices of the same size. Show that if B is not invertible then AB is not invertible.

Section 2.2 Applications of Determinant

Cramer's Rule

Suppose one wants to solve the system $A\mathbf{x} = \mathbf{b}$ and suppose A is $n \times n$ with $|A| \neq 0$. We already know that since $|A| \neq 0$, A is invertible and so a unique solution exists. Cramer's Rule gives a way of expressing the solution in terms of determinants. Let A_i stand for the matrix obtained by replacing column i of A by the column vector $\mathbf{b}$. Then we have $\mathbf{x} = (x_1, x_2, \cdots x_n)$ is a solution where

$$x_1 = \frac{|A_1|}{|A|}$$

$$x_2 = \frac{|A_2|}{|A|}$$

$$\vdots$$

$$x_n = \frac{|A_n|}{|A|}.$$

Example *Consider the following system.*

$$\begin{aligned} 2x + y + z &= 7 \\ x + 2y + z &= 8 \\ x + y + 2z &= 9 \end{aligned}.$$

Suppose you want to solve for z without solving for the other unknowns. Using Cramer's Rule we have

$$A = \begin{bmatrix} 2 & 1 & 1 \\ 1 & 2 & 1 \\ 1 & 1 & 2 \end{bmatrix}$$

and

$$A_3 = \begin{bmatrix} 2 & 1 & 7 \\ 1 & 2 & 8 \\ 1 & 1 & 9 \end{bmatrix},$$

so that

$$z = \frac{|A_3|}{|A|} = \frac{12}{4} = 3.$$

Each unknown equals the quotient of two determinants. The denominator is the determinant of the coefficient matrix. If we wish to solve for y in the system above we have

$$A_2 = \begin{bmatrix} 2 & 7 & 1 \\ 1 & 8 & 1 \\ 1 & 9 & 2 \end{bmatrix}$$

and

$$y = \frac{|A_2|}{|A|} = \frac{8}{4} = 2.$$

Vandermonde Determinant

One can calculate that

$$\begin{vmatrix} 1 & a & a^2 \\ 1 & b & b^2 \\ 1 & c & c^2 \end{vmatrix} = (b-a)(c-a)(c-b).$$

It can be shown that

$$\begin{vmatrix} 1 & x_1 & x_1^2 & x_1^3 & \cdots & x_1^{n-1} \\ 1 & x_2 & x_2^2 & x_2^3 & \cdots & x_2^{n-1} \\ \vdots & \vdots & \vdots & \vdots & & \vdots \\ 1 & x_n & x_n^2 & x_n^3 & \cdots & x_n^{n-1} \end{vmatrix} = \prod_{j>i}(x_j - x_i)$$

Determinants

A matrix of this form is called a Vandermonde matrix. The formula says that the determinant is the product of all differences $x_j - x_i$ where $j > i$. Thus we see that if the x_i are all distinct ($x_i \neq x_j$ for $i \neq j$), the determinant is not zero.

Example *Calulate the determinant*

$$\begin{vmatrix} 1 & -1 & 1 & -1 & 1 \\ 1 & 1 & 1 & 1 & 1 \\ 1 & 2 & 4 & 8 & 16 \\ 1 & -2 & 4 & -8 & 16 \\ 1 & 3 & 9 & 27 & 81 \end{vmatrix}.$$

This matrix is of the Vandermonde form with $x_1 = -1$, $x_2 = 1$, $x_3 = 2$, $x_4 = -2$, and $x_5 = 3$. Thus using the formula we know the determinant equals

$$(3+2)(3-2)(3-1)(3+1)(-2-2)(-2-1)(-2+1)(2-1)(2+1)(1+1)$$

which is

$$(5)(1)(2)(4)(-4)(-3)(-1)(1)(3)(2) = (40)(-72) = -2880.$$

Application of Vandermonde matrices

Fact *Given a set of n points in the plane $(x_1, y_1), (x_2, y_2), \cdots, (x_n, y_n)$ such that no two of the points lie on the same vertical line (no two of the points have the same x coordinate) then there exists a unique polynomial of degree $n-1$ or less $a_0 + a_1 x + a_2 x^2 + \cdots + a_{n-1} x^{n-1} = y$ which passes through all these points.*

Reason. If the polynomial is to pass through all the points, we must have

$$a_0 + a_1 x_1 + a_2 x_1^2 + \cdots + a_{n-1} x_1^{n-1} = y_1$$
$$a_0 + a_1 x_2 + a_2 x_2^2 + \cdots + a_{n-1} x_2^{n-1} = y_2$$

$$a_0 + a_1 x_n + a_2 x_n^2 + \cdots + a_{n-1} x_n^{n-1} = y_n.$$

This can be viewed as a system of n linear equations in the unknowns $\{a_0, a_1, \cdots, a_{n-1}\}$. We can write it in matrix form as

$$\begin{bmatrix} 1 & x_1 & x_1^2 & x_1^3 & \cdots & x_1^{n-1} \\ 1 & x_2 & x_2^2 & x_2^3 & \cdots & x_2^{n-1} \\ \vdots & \vdots & \vdots & \vdots & & \vdots \\ 1 & x_n & x_n^2 & x_n^3 & \cdots & x_n^{n-1} \end{bmatrix} \begin{bmatrix} a_0 \\ a_1 \\ \vdots \\ a_{n-1} \end{bmatrix} = \begin{bmatrix} y_1 \\ y_2 \\ \vdots \\ y_n \end{bmatrix}.$$

The coefficient matrix is a Vandermonde matrix and its determinant is not zero because of the condition placed on the x_i. Thus there is a unique solution for the a_i.

Exercises

1. Show that there exists a unique parabola that passes through the points $(0, 2)$, $(2, 10)$, and $(-2, -2)$ by determining its equation.

2. Find numbers a_0, a_1, a_2, a_3 such that the equation $y = a_0 + a_1 x + a_2 x^2 + a_3 x^3$ passes through the points $(-1, 2)$, $(0, 4)$, $(1, 10)$, and $(2, 26)$.

3. Use Cramer's Rule to solve the following system for for x only.
$$\begin{array}{rcrcrcr} -2x & + & 3y & - & z & = & 1 \\ x & + & 2y & - & z & = & 4 \\ -2x & - & y & + & z & = & -3 \end{array}.$$

4. Find the complete solution to the system in problem 3 by finding the RREF of the augmented matrix of the system.

5. Solve the following system for w using Cramer's Rule.
$$\begin{array}{rcrcrcrcr} 4w & + & 3x & + & 2y & + & z & = & 10 \\ -3w & + & 8x & - & 3y & & & = & 2 \\ w & + & x & + & y & + & z & = & 4 \\ w & - & x & - & y & + & 2z & = & 1 \end{array}.$$

6. Solve the system in problem 5 for all the unknowns by inverting the coefficient matrix. (Use a computer program if possible.)

Chapter 3

Diagonalization

Section 3.1 Eigenvalues and Eigenvectors

A **linear transformation** from R^n to R^n is a function T defined on all vectors in R^n and taking values in R^n which satisfies the following two conditions. For all vectors $\mathbf{u}$ and $\mathbf{v}$ in R^n and all scalars c,
   i) $T(\mathbf{u} + \mathbf{v}) = T(\mathbf{u}) + T(\mathbf{v})$
   ii) $T(c\mathbf{u}) = cT(\mathbf{u})$.

Example *The mapping defined by $T(\mathbf{v}) = A\mathbf{v}$, where A is an $n \times n$ matrix and $\mathbf{v}$ is a vector of R^n written in column form is a linear transformation from R^n to R^n.*

Reason. $T(\mathbf{v} + \mathbf{u}) = A(\mathbf{v} + \mathbf{u}) = A\mathbf{v} + A\mathbf{u} = T(\mathbf{v}) + T(\mathbf{u})$ by the distributive law of matrix multiplication, and $T(c\mathbf{v}) = A(c\mathbf{v}) = cA\mathbf{v} = cT(\mathbf{v})$, because of the associative law of matrix multiplication and the fact that scalars commute with matrices and vectors.

Not only is sending $\mathbf{v}$ to $A\mathbf{v}$ a linear transformation, but every linear transformation from R^n to R^n is of this form. To see this let

$$\mathbf{e}_1 = \begin{bmatrix} 1 \\ 0 \\ 0 \\ \vdots \\ 0 \end{bmatrix}, \quad \mathbf{e}_2 = \begin{bmatrix} 0 \\ 1 \\ 0 \\ \vdots \\ 0 \end{bmatrix}, \quad \cdots \quad \mathbf{e}_n = \begin{bmatrix} 0 \\ 0 \\ \vdots \\ 0 \\ 1 \end{bmatrix}.$$

This is the standard basis of R^n in column form. Let T be an arbitrary linear transformation. Let $T(\mathbf{e}_i) = \mathbf{v}_i$. Then

$$\begin{bmatrix} c_1 \\ c_2 \\ \vdots \\ c_n \end{bmatrix} = c_1 \mathbf{e}_1 + c_2 \mathbf{e}_2 + \cdots + c_n \mathbf{e}_n$$

$$\begin{aligned} T(c_1 \mathbf{e}_1 + c_2 \mathbf{e}_2 + \cdots + c_n \mathbf{e}_n) &= c_1 T(\mathbf{e}_1) + c_2 T(\mathbf{e}_2) + \cdots + c_n T(\mathbf{e}_n) \\ &= c_1 \mathbf{v}_1 + c_2 \mathbf{v}_2 + \cdots + c_n \mathbf{v}_n. \end{aligned}$$

In matrix form

$$c_1 \mathbf{v}_1 + c_2 \mathbf{v}_2 + \cdots + c_n \mathbf{v}_n = \begin{bmatrix} \mathbf{v}_1 & \mathbf{v}_2 & \cdots & \mathbf{v}_n \end{bmatrix} \begin{bmatrix} c_1 \\ c_2 \\ \vdots \\ c_n \end{bmatrix},$$

thus

$$T \begin{bmatrix} c_1 \\ c_2 \\ \vdots \\ c_n \end{bmatrix} = \begin{bmatrix} \mathbf{v}_1 & \mathbf{v}_2 & \cdots & \mathbf{v}_n \end{bmatrix} \begin{bmatrix} c_1 \\ c_2 \\ \vdots \\ c_n \end{bmatrix}.$$

Let

$$A = \begin{bmatrix} \mathbf{v}_1 & \mathbf{v}_2 & \cdots & \mathbf{v}_n \end{bmatrix},$$

then we have

$$T \begin{bmatrix} c_1 \\ c_2 \\ \vdots \\ c_n \end{bmatrix} = A \begin{bmatrix} c_1 \\ c_2 \\ \vdots \\ c_n \end{bmatrix}.$$

We have shown the following

Diagonalization

Fact *An arbitrary linear transformation T from R^n to R^n can be obtained by matrix multiplication on the left. The matrix A that works for T has as its columns the images under T of the standard basis vectors of R^n.*

Given a linear transformation T, we say k is an **eigenvalue** of T if there exists a nonzero vector $\mathbf{v}$ such that $T(\mathbf{v}) = k\mathbf{v}$. The vector $\mathbf{v}$ is called an **eigenvector** of T.

In order to find all eigenvalues and eigenvectors of T we need to solve the matrix equation $A\mathbf{v} = k\mathbf{v}$. This is equivalent to solving $(A - kI)\mathbf{v} = \mathbf{O}$. Thus the eigenvalues of T are the numbers k for which the matrix $(A - kI)$ has a nonzero null space. If $T(\mathbf{v}) = A\mathbf{v}$, the eigenvalues and eigenvectors of T are also called eigenvalues and eigenvectors of A.

Fact *The number k is an eigenvalue of A if and only if $|A - kI| = 0$.*

Letting k be a variable, $|A - kI|$ is a polynomial of degree n in k, called the **characteristic polynomial** of A, and $|A - kI| = 0$ is an equation in k, which is called the **characteristic equation** of A. For each real root r of this equation (eigenvalue) the eigenvectors of A for the eigenvalue r are the nonzero vectors in the null space of $(A - rI)$, which is called the **eigenspace** of A for r.

Procedure *Find the characteristic equation of A by subtracting the letter k from the diagonal entries of A, taking the determinant of this matrix and setting it equal to zero. The roots of this equation are the eigenvalues of A. For each of these roots r, the eigenvectors of A for r are the nonzero vectors in the null space of $(A - rI)$.*

Example Let
$$A = \begin{bmatrix} 1 & 2 \\ -1 & 4 \end{bmatrix}.$$

Then
$$A - kI = \begin{bmatrix} 1-k & 2 \\ -1 & 4-k \end{bmatrix}.$$

$$|A - kI| = \begin{vmatrix} 1-k & 2 \\ -1 & 4-k \end{vmatrix}$$
$$= k^2 - 5k + 4 + 2$$
$$= k^2 - 5k + 6$$
$$= (k-2)(k-3)$$

We solve
$$(k-2)(k-3) = 0.$$
Thus the eigenvalues are 2 and 3.

$$A - 2I = \begin{bmatrix} -1 & 2 \\ -1 & 2 \end{bmatrix}.$$

The rank of this matrix is 1, hence the dimension of the nullspace is 1. It is clear by inspection that

$$\begin{bmatrix} -1 & 2 \\ -1 & 2 \end{bmatrix} \begin{bmatrix} 2 \\ 1 \end{bmatrix} = \begin{bmatrix} 0 \\ 0 \end{bmatrix}.$$

Thus
$$\begin{bmatrix} 2 \\ 1 \end{bmatrix}$$
is a basis of the null space, and the eigenvectors of A for the eigenvalue 2 are all vectors of the form

$$\begin{bmatrix} 2t \\ t \end{bmatrix},$$

where t is any nonzero number. Note that

$$\begin{bmatrix} 1 & 2 \\ -1 & 4 \end{bmatrix} \begin{bmatrix} 2t \\ t \end{bmatrix} = \begin{bmatrix} 4t \\ 2t \end{bmatrix} = 2 \begin{bmatrix} 2t \\ t \end{bmatrix}.$$

Now we look for the eigenvectors of A for the eigenvalue 3.

$$A - 3I = \begin{bmatrix} -2 & 2 \\ -1 & 1 \end{bmatrix},$$

and by inspection

$$\begin{bmatrix} -2 & 2 \\ -1 & 1 \end{bmatrix} \begin{bmatrix} 1 \\ 1 \end{bmatrix} = \begin{bmatrix} 0 \\ 0 \end{bmatrix},$$

Diagonalization

thus

$$\begin{bmatrix} 1 \\ 1 \end{bmatrix}$$

is a basis of the eigenspace of A for the eigenvalue 3. We can check by computing

$$\begin{bmatrix} 1 & 2 \\ -1 & 4 \end{bmatrix} \begin{bmatrix} 1 \\ 1 \end{bmatrix} = \begin{bmatrix} 3 \\ 3 \end{bmatrix} = 3 \begin{bmatrix} 1 \\ 1 \end{bmatrix}.$$

For $n = 2$, the characteristic equation is a quadratic. If you have trouble factoring, you can use the quadratic formula. This will give the roots (which may be complex in which case A has no real eigenvalues.) For $n > 2$, finding the roots may be difficult. In this case we may have to settle for approximations obtained with computer assistance.

Fact *If A is $n \times n$, the dimension of the eigenspace of A for an eigenvalue r of A equals $n - rank(A - rI)$.*

Reason. This eigenspace equals the null space of $(A - rI)$.

Example *Consider*

$$A = \begin{bmatrix} 2 & 0 & 0 & 0 \\ 0 & 2 & 0 & 0 \\ 0 & 0 & 3 & -1 \\ 0 & 0 & 1 & 1 \end{bmatrix}.$$

Then

$$|A - kI| = \begin{vmatrix} 2-k & 0 & 0 & 0 \\ 0 & 2-k & 0 & 0 \\ 0 & 0 & 3-k & -1 \\ 0 & 0 & 1 & 1-k \end{vmatrix}$$

$$= (2-k) \begin{vmatrix} 2-k & 0 & 0 \\ 0 & 3-k & -1 \\ 0 & 1 & 1-k \end{vmatrix}$$

$$= (2-k)^2 \begin{vmatrix} 3-k & -1 \\ 1 & 1-k \end{vmatrix}$$

$$= (2-k)^2(k^2 - 4k + 4)$$

$$= (k-2)^4.$$

Thus A has just one eigenvalue, namely 2. To find the eigenvectors of A we find the null space of $(A - 2I)$.

$$A - 2I = \begin{bmatrix} 0 & 0 & 0 & 0 \\ 0 & 0 & 0 & 0 \\ 0 & 0 & 1 & -1 \\ 0 & 0 & 1 & -1 \end{bmatrix}.$$

Since this is a matrix of rank 1, the null space has dimension 3. We find a basis by finding

$$H = H(A - 2I) = \begin{bmatrix} 0 & 0 & 0 & 0 \\ 0 & 0 & 0 & 0 \\ 0 & 0 & 1 & -1 \\ 0 & 0 & 0 & 0 \end{bmatrix},$$

and then

$$H - I = \begin{bmatrix} -1 & 0 & 0 & 0 \\ 0 & -1 & 0 & 0 \\ 0 & 0 & 0 & -1 \\ 0 & 0 & 0 & -1 \end{bmatrix}.$$

The three nonzero columns of this matrix are a basis of the eigenspace of A for the eigenvalue 2. The eigenvectors of A for $k = 2$ are all nonzero linear combinations of $(-1, 0, 0, 0)$, $(0, -1, 0, 0)$, and $(0, 0, -1, -1)$. In other words all nonzero vectors of the form (r, s, t, t). To check this we compute

$$A \begin{bmatrix} r \\ s \\ t \\ t \end{bmatrix} = \begin{bmatrix} 2 & 0 & 0 & 0 \\ 0 & 2 & 0 & 0 \\ 0 & 0 & 3 & -1 \\ 0 & 0 & 1 & 1 \end{bmatrix} \begin{bmatrix} r \\ s \\ t \\ t \end{bmatrix} = \begin{bmatrix} 2r \\ 2s \\ 2t \\ 2t \end{bmatrix} = 2 \begin{bmatrix} r \\ s \\ t \\ t \end{bmatrix}.$$

The **algebraic multiplicity** of an eigenvalue r is its multiplicity as a root of the characteristic equation (highest power of $(k - r)$ that is a factor of the characteristic polynomial). For example the algebraic multiplicity of the eigenvalue 2 in the preceding example is 4. The **geometric multiplicity** of an eigenvalue is the dimension of its eigenspace. In the preceding example the geometric multiplicity of the eigenvalue 2 is 3.

Diagonalization

Fact For every eigenvalue k, geometric multiplicity $\leq$ algebraic multiplicity.

Fact If the characteristic polynomial of an $n \times n$ matrix has n distinct eigenvalues then all algebraic and geometric multiplicities equal one.

Example Let
$$A = \begin{bmatrix} 2 & 2 & 1 \\ 2 & -1 & -2 \\ 1 & -2 & 2 \end{bmatrix}.$$

Then
$$|A - kI| = \begin{vmatrix} 2-k & 2 & 1 \\ 2 & -1-k & -2 \\ 1 & -2 & 2-k \end{vmatrix}.$$

Adding row three to row one gives
$$\begin{vmatrix} 3-k & 0 & 3-k \\ 2 & -1-k & -2 \\ 1 & -2 & 2-k \end{vmatrix} = (3-k) \begin{vmatrix} 1 & 0 & 1 \\ 2 & -1-k & -2 \\ 1 & -2 & 2-k \end{vmatrix}.$$

Subtracting column one from column three and then expanding by minors around row one gives
$$(3-k) \begin{vmatrix} 1 & 0 & 0 \\ 2 & -1-k & -4 \\ 1 & -2 & 1-k \end{vmatrix} = (3-k) \begin{vmatrix} -1-k & -4 \\ -2 & 1-k \end{vmatrix}$$
$$= (3-k)(-1+k^2-8)$$
$$= (3-k)(k^2-9)$$
$$= (3-k)(k-3)(k+3)$$
$$= -(k-3)^2(k+3).$$

Thus we see that the eigenvalues of A are 3 and -3. The algebraic multiplicity of -3 is one, thus without further calculation we know that the geometric multiplicity of -3 is also one. The algebraic multiplicity of 3 is two, so we know that the geometric multiplicity of 3 is either one or two. To see which it is we need only find the rank of the matrix $(A - 3I)$.

$$A - 3I = \begin{bmatrix} -1 & 2 & 1 \\ 2 & -4 & -2 \\ 1 & -2 & -1 \end{bmatrix}.$$

Since the second and third rows are just multiples of the first row, we see that this matrix has rank one, so that the geometric multiplicity of 3 is two. To find a basis for this eigenspace, find

$$RREF(A - 3I) = \begin{bmatrix} 1 & -2 & -1 \\ 0 & 0 & 0 \\ 0 & 0 & 0 \end{bmatrix},$$

which is already the Hermite form H, and forming

$$H - I = \begin{bmatrix} 0 & -2 & -1 \\ 0 & -1 & 0 \\ 0 & 0 & -1 \end{bmatrix}$$

we get as a basis of the eigenspace for $k = 3$ the vectors

$$\begin{bmatrix} -2 \\ -1 \\ 0 \end{bmatrix} \text{ and } \begin{bmatrix} -1 \\ 0 \\ -1 \end{bmatrix}.$$

We leave the computation of the eigenvectors for $k = -3$ to the reader.

Warning While doing row and/or column operations on the matrix $(A - kI)$ can help find the characteristic equation as in the above example, doing row and/or column operations on the matrix A may (usually will) change the eigenvalues. Thus when finding eigenvalues do not do row or column operations until after forming $A - kI$.

Exercises

1. For each of the following matrices calculate the eigenvalues and a basis of the corresponding eigenspaces.

   (a) $\begin{bmatrix} 2 & 3 \\ 3 & 2 \end{bmatrix}$

   (b) $\begin{bmatrix} 1 & 4 \\ 3 & 2 \end{bmatrix}$

   (c) $\begin{bmatrix} 3 & -6 & 8 \\ 4 & -7 & 8 \\ 2 & -6 & 9 \end{bmatrix}$

Diagonalization

(d) $\begin{bmatrix} 1 & 1 & 0 \\ -1 & 3 & 0 \\ 0 & 0 & 2 \end{bmatrix}$

(e) $\begin{bmatrix} -1 & -1 & 0 \\ 4 & 3 & 0 \\ 0 & 0 & 2 \end{bmatrix}$

(f) $\begin{bmatrix} 5 & -3 & 3 \\ -3 & 5 & -3 \\ -9 & 9 & -7 \end{bmatrix}$

(g) $\begin{bmatrix} 2 & 1 & 0 & 0 \\ 2 & 3 & 0 & 0 \\ 0 & 0 & 1 & 2 \\ 0 & 0 & -4 & 7 \end{bmatrix}$.

2. For each of the matrices in problem 1 find the geometric and algebraic multiplicities of each eigenvalue.

3. If 0 is an eigenvalue of a matrix, show that the columns of the matrix must be linearly dependent.

4. Prove the converse of the implication proved in problem 3.

Section 3.2 Diagonalization

A square matrix A is said to be **diagonalizable** in case there exists an invertible matrix P such that $P^{-1}AP = D$, where D is a matrix that has zeros everywhere except on the main diagonal. D is called a **diagonal matrix**. For example

$$D = \begin{bmatrix} 3 & 0 & 0 & 0 \\ 0 & 2 & 0 & 0 \\ 0 & 0 & -1 & 0 \\ 0 & 0 & 0 & 12 \end{bmatrix}$$

is a 4×4 diagonal matrix.

Fact *If $P^{-1}AP = D$ and D is a diagonal matrix, then the diagonal entries of D are the eigenvalues of A.*

Reason. Since
$$P^{-1}AP = D,$$
we have
$$P^{-1}(A - kI)P = D - kI.$$
Taking determinants gives
$$\left|P^{-1}\right| |A - kI| |P| = |D - kI|.$$
Since $|P^{-1}||P| = 1$, this gives
$$\begin{aligned}
|A - kI| &= |D - kI| \\
&= \begin{vmatrix} d_1 - k & 0 & 0 & \cdots & 0 \\ 0 & d_2 - k & 0 & \cdots & 0 \\ 0 & 0 & \ddots & & \vdots \\ \vdots & \vdots & & \ddots & 0 \\ 0 & 0 & \cdots & 0 & d_n - k \end{vmatrix} \\
&= (d_1 - k)(d_2 - k) \cdots (d_n - k).
\end{aligned}$$

Fact *A necessary and sufficient condition for the $n \times n$ matrix A to be diagonalizable is that the eigenvalues of A are all real and that there exists a basis of R^n consisting entirely of eigenvectors of A.*

Fact *If $\{v_1, v_2, \cdots v_n\}$ is a basis of R^n consisting entirely of eigenvectors of A, and if P is the matrix whose columns are these eigenvectors, $P = \begin{bmatrix} v_1 & v_2 & \cdots & v_n \end{bmatrix}$, then $P^{-1}AP = D$.*

Reason. Suppose $v_1, v_2, \cdots v_n$ and P are as above, and let d_i be the eigenvalue of A that goes with the eigenvector v_i. Then $Av_i = d_i v_i$ for $i = 1, 2, \cdots, n$, and we have
$$\begin{aligned}
AP &= A \begin{bmatrix} v_1 & v_2 & \cdots & v_n \end{bmatrix} \\
&= \begin{bmatrix} Av_1 & Av_2 & \cdots & Av_n \end{bmatrix} \\
&= \begin{bmatrix} d_1 v_1 & d_2 v_2 & \cdots & d_n v_n \end{bmatrix},
\end{aligned}$$

Diagonalization

and

$$[d_1\mathbf{v}_1 \ d_2\mathbf{v}_2 \ \cdots \ d_n\mathbf{v}_n] = [\mathbf{v}_1 \ \mathbf{v}_2 \ \cdots \ \mathbf{v}_n]\begin{bmatrix} d_1 & 0 & \cdots & 0 \\ 0 & d_2 & & \vdots \\ \vdots & & \ddots & 0 \\ 0 & 0 & \cdots & d_n \end{bmatrix}$$

$$= PD.$$

Thus we have
$$AP = PD,$$
and multiplying both sides on the left by P^{-1} gives
$$P^{-1}AP = D.$$

Fact *If $P^{-1}AP = D$, where D is a diagonal matrix, then the columns of P are a basis of R^n consisting entirely of eigenvectors.*

Reason. From the comutations above we see that $AP = PD$ is equivalent to stating that the columns of P are eigenvectors of A. Since P is assumed invertible, the columns are a basis of R^n.

Remark *The preceding two facts together show that an $n \times n$ matrix A is diagonalizable if and only if there exists a basis of R^n consisting entirely of eigenvectors of A.*

Example *The matrix*
$$\begin{bmatrix} 0 & 1 \\ -1 & 0 \end{bmatrix}$$
is not diagonalizable because the characteristic polynomial of this matrix equals
$$\begin{vmatrix} -k & 1 \\ -1 & -k \end{vmatrix} = k^2 + 1,$$
and this polynomial has no real roots.

Example *The matrix*
$$\begin{bmatrix} 2 & 0 \\ 1 & 2 \end{bmatrix}$$

is not diagonalizable. This time the problem is not with the eigenvalues. We have
$$\begin{vmatrix} 2-k & 0 \\ 1 & 2-k \end{vmatrix} = (2-k)^2.$$
Thus 2 is the only eigenvalue. The eigenspace for 2 is the null space of the matrix
$$\begin{bmatrix} 0 & 0 \\ 1 & 0 \end{bmatrix}.$$
But this eigenspace has dimension one, and thus does not contain a basis of R^2.

Fact If you find a basis for each eigenspace of A, the union of these vectors will be linearly independent.

Fact An $n \times n$ matrix A is diagonalizable if and only if all its eigenvalues are real and for every eigenvalue r of A the algebraic and geometric multiplicities are equal.

Fact If an $n \times n$ matrix A has n distinct real eigenvalues, then A is diagonalizable.

Warning The matrix A may be diagonalizable even though it has repeated roots. To tell if it is or not it is only necessary to find the geometric multiplicities of the repeated roots and compare to the algebraic multiplicites.

Example Let
$$A = \begin{bmatrix} 2 & 2 & 1 \\ 2 & -1 & -2 \\ 1 & -2 & 2 \end{bmatrix}.$$
In the previous section this matrix was an example and we found its characteristic polynomial equal to $(k-3)^2(k+3)$. We need to know the dimension of the eigenspace for $k = 3$.
$$A - 3I = \begin{bmatrix} -1 & 2 & 1 \\ 2 & -4 & -2 \\ 1 & -2 & -1 \end{bmatrix}$$

Diagonalization

has rank one, thus the dimension of this eigenspace (null space of $(A-3I)$) is two. Geometric multiplicity equals algebraic multiplicity for $k=3$. We do not need to check the geometric multiplicity of -3 because we know there is at least one nonzero eigenvector for -3 (since $|A+3I|=0$), and one is all we need (algebraic multiplicity of -3 is one).

Now that we know A is diagonalizable, we continue with this example to find an invertible matrix P such that $P^{-1}AP$ is diagonal. We need a basis for each eigenspace.

$$H(A-3I) = \begin{bmatrix} 1 & -2 & -1 \\ 0 & 0 & 0 \\ 0 & 0 & 0 \end{bmatrix},$$

so a basis for the eigenspace for $k=3$ is

$$\begin{bmatrix} -2 \\ -1 \\ 0 \end{bmatrix}, \text{ and } \begin{bmatrix} -1 \\ 0 \\ -1 \end{bmatrix}.$$

For $k=-3$, we have

$$A+3I = \begin{bmatrix} 5 & 2 & 1 \\ 2 & 2 & -2 \\ 1 & -2 & 5 \end{bmatrix},$$

$$RREF(A+3I) = \begin{bmatrix} 1 & 0 & 1 \\ 0 & 1 & -2 \\ 0 & 0 & 0 \end{bmatrix} = H(A+3I),$$

thus

$$\begin{bmatrix} 1 \\ -2 \\ -1 \end{bmatrix}$$

is a basis of the eigenspace for $k=-3$. Let

$$P = \begin{bmatrix} -2 & -1 & 1 \\ -1 & 0 & -2 \\ 0 & -1 & -1 \end{bmatrix},$$

then it must follow that

$$P^{-1}AP = \begin{bmatrix} 3 & 0 & 0 \\ 0 & 3 & 0 \\ 0 & 0 & -3 \end{bmatrix} = D.$$

To check this it is easier to check that P is invertible (by checking that it has a nonzero determinant or that its RREF is I) and then check that $AP = PD$. (You don't have to find P^{-1} to perform this check.) This is only a check on one's work, since we know that if the columns of P are linearly independent eigenvectors of A, then $P^{-1}AP$ must have the desired form. We compute

$$AP = \begin{bmatrix} 2 & 2 & 1 \\ 2 & -1 & -2 \\ 1 & -2 & 2 \end{bmatrix} \begin{bmatrix} -2 & -1 & 1 \\ -1 & 0 & -2 \\ 0 & -1 & -1 \end{bmatrix} = \begin{bmatrix} -6 & -3 & -3 \\ -3 & 0 & 6 \\ 0 & -3 & 3 \end{bmatrix},$$

and

$$PD = \begin{bmatrix} -2 & -1 & 1 \\ -1 & 0 & -2 \\ 0 & -1 & -1 \end{bmatrix} \begin{bmatrix} 3 & 0 & 0 \\ 0 & 3 & 0 \\ 0 & 0 & -3 \end{bmatrix} = \begin{bmatrix} -6 & -3 & -3 \\ -3 & 0 & 6 \\ 0 & -3 & 3 \end{bmatrix}$$

so we have
$$AP = PD.$$

Expanding by minors around the middle column of P gives

$$\begin{aligned} |P| &= \begin{vmatrix} -2 & -1 & 1 \\ -1 & 0 & -2 \\ 0 & -1 & -1 \end{vmatrix} \\ &= 1 \begin{vmatrix} -1 & -2 \\ 0 & -1 \end{vmatrix} + 1 \begin{vmatrix} -2 & 1 \\ -1 & -2 \end{vmatrix} \\ &= 1 + 4 + 5 = 10. \end{aligned}$$

Thus P is invertible. This completes the check.

Exercises

1. For each of the following matrices, determine whether or not the matrix is diagonalizable.

Diagonalization

(a) $\begin{bmatrix} 2 & 3 \\ 5 & 4 \end{bmatrix}$

(b) $\begin{bmatrix} 0 & 1 \\ 1 & 1 \end{bmatrix}$

(c) $\begin{bmatrix} 1 & 1 \\ 1 & 1 \end{bmatrix}$

(d) $\begin{bmatrix} 7 & 4 & -4 \\ 4 & 7 & -4 \\ -1 & -1 & 4 \end{bmatrix}$

(e) $\begin{bmatrix} 2 & 2 & 1 \\ 2 & -1 & -2 \\ 1 & -2 & 2 \end{bmatrix}$

(f) $\begin{bmatrix} 3 & 0 & 0 & 0 \\ 1 & 3 & 0 & 0 \\ -1 & 1 & 2 & 0 \\ 1 & -1 & 1 & 3 \end{bmatrix}$

(g) $\begin{bmatrix} 1 & 1 & 0 & 0 \\ 1 & 1 & 0 & 0 \\ 0 & 0 & 2 & 0 \\ 0 & 0 & 1 & 3 \end{bmatrix}$

(h) $\begin{bmatrix} -3 & 1 & 1 & 1 \\ 1 & -3 & 1 & 1 \\ 1 & 1 & -3 & 1 \\ 1 & 1 & 1 & -3 \end{bmatrix}.$

2. For each of the matrices A in Exercise 1 which are diagonalizable, find a matrix P which diagonalizes A (such that $P^{-1}AP$ is diagonal).

Section 3.3 Orthogonal Basis

If $\mathbf{u} = (a_1, a_2, \cdots, a_n)$ and $\mathbf{v} = (b_1, b_2, \cdots, b_n)$ are two vectors in R^n, the **dot product** of $\mathbf{u}$ and $\mathbf{v}$ (written $\mathbf{u} \cdot \mathbf{v}$) is defined by

$$\mathbf{u} \cdot \mathbf{v} = a_1 b_1 + a_2 b_2 + \cdots + a_n b_n.$$

Notice that the dot product of two vectors is a scalar. Also note that if $\mathbf{u}$ is written as a row and $\mathbf{v}$ as a column, the dot product can be expressed in matrix notation

$$\mathbf{u} \cdot \mathbf{v} = \begin{bmatrix} a_1 & a_2 & \cdots & a_n \end{bmatrix} \begin{bmatrix} b_1 \\ b_2 \\ \vdots \\ b_n \end{bmatrix}.$$

Fact *We can describe matrix multiplication in terms of the dot product: in the product AB of two $n \times n$ matrices, the entry in row i column j is the dot product of row i of A and column j of B.*

Two vectors $\mathbf{u}$ and $\mathbf{v}$ in R^n are said to be **orthogonal** if $\mathbf{u} \cdot \mathbf{v} = 0$.

Fact *A vector in the null space of A is orthogonal to all the rows of A.*

Reason. Let A have rows $\mathbf{v}_1, \mathbf{v}_2, \cdots, \mathbf{v}_n$. Then if $\mathbf{u}$ is a vector in the null space of A ($\mathbf{u}$ written as a column),

$$A\mathbf{u} = \begin{bmatrix} \mathbf{v}_1 \\ \mathbf{v}_2 \\ \vdots \\ \mathbf{v}_n \end{bmatrix} \mathbf{u} = \begin{bmatrix} \mathbf{v}_1 \cdot \mathbf{u} \\ \mathbf{v}_2 \cdot \mathbf{u} \\ \vdots \\ \mathbf{v}_n \cdot \mathbf{u} \end{bmatrix} = \begin{bmatrix} 0 \\ 0 \\ \vdots \\ 0 \end{bmatrix}.$$

A set of vectors $\{\mathbf{v}_1, \mathbf{v}_2, \cdots, \mathbf{v}_k\}$ is an **orthogonal set** if the vectors are pairwise orthogonal, in other words if $\mathbf{v}_i \cdot \mathbf{v}_j = 0$ for all $i \neq j$.

Fact *An orthogonal set of nonzero vectors is linearly independent.*

Diagonalization

Reason. Let $v_1, v_2, \cdots, v_k$ be an orthogonal set of nonzero vectors. If

$$a_1 v_1 + a_2 v_2 + \cdots + a_k v_k = O,$$

then taking the dot product of v_1 with each side of the equation gives

$$a_1(v_1 \cdot v_1) = 0,$$

from which it follow that $a_1 = 0$. Similarly taking the dot product of v_i with both sides shows that $a_i = 0$, so $v_1, v_2, \cdots, v_k$ are linearly independent.

An **orthogonal basis** for a subspace W is a basis of W that is an orthogonal set.

Given a nonzero subspace W of R^n we want to find an orthogonal basis of W. In this section we present a solution to this problem.

Procedure (1) *The technique depends on how W is defined. First assume W is defined as the null space of an $n \times n$ matrix A. Then we find $RREF(A)$, and from this delete the zero rows. Call this matrix B. W is still the null space of B. Let k be the number of rows of B, then k is the rank of A, and $n - k$ is the dimension of W. Let $v_1, v_2, \cdots, v_k$ be the rows of B, so that*

$$B = \begin{bmatrix} v_1 \\ v_2 \\ \vdots \\ v_k \end{bmatrix}.$$

Next find a nonzero vector w_{k+1} in W. (If necessary use $H(A) - I$, the columns of which are in W.) Let

$$B' = \begin{bmatrix} v_1 \\ v_2 \\ \vdots \\ v_k \\ w_{k+1} \end{bmatrix}.$$

Find a vector in the null space of B', call it $\mathbf{w}_{k+2}$. This vector will be in W and also orthogonal to $\mathbf{w}_{k+1}$. Adjoin the new vector to B' as another row, forming

$$B'' = \begin{bmatrix} \mathbf{v}_1 \\ \mathbf{v}_2 \\ \vdots \\ \mathbf{v}_k \\ \mathbf{w}_{k+1} \\ \mathbf{w}_{k+2} \end{bmatrix}.$$

Find a vector $\mathbf{w}_{k+3}$ in the null space of B''. This vector will be in W and also orthogonal to both $\mathbf{w}_{k+1}$ and $\mathbf{w}_{k+2}$. Continue in this way until there are n rows in the matrix. The set $\{\mathbf{w}_{k+1}, \mathbf{w}_{k+2}, \cdots, \mathbf{w}_n\}$ will be an orthogonal basis of W.

Example *Find an orthogonal basis of W = the null space of A, where*

$$A = \begin{bmatrix} 2 & 2 & 2 & 10 \\ 1 & 1 & 1 & 5 \\ 3 & 3 & -1 & 7 \\ 4 & 4 & 3 & 18 \end{bmatrix}.$$

Then

$$RREF(A) = \begin{bmatrix} 1 & 1 & 0 & 3 \\ 0 & 0 & 1 & 2 \\ 0 & 0 & 0 & 0 \\ 0 & 0 & 0 & 0 \end{bmatrix},$$

so we let

$$B = \begin{bmatrix} 1 & 1 & 0 & 3 \\ 0 & 0 & 1 & 2 \end{bmatrix},$$

and W is also the null space of B.
The Hermite form of A is

$$H(A) = \begin{bmatrix} 1 & 1 & 0 & 3 \\ 0 & 0 & 0 & 0 \\ 0 & 0 & 1 & 2 \\ 0 & 0 & 0 & 0 \end{bmatrix},$$

Diagonalization

and
$$H(A) - I = \begin{bmatrix} 0 & 1 & 0 & 3 \\ 0 & -1 & 0 & 0 \\ 0 & 0 & 0 & 2 \\ 0 & 0 & 0 & -1 \end{bmatrix}.$$

Thus
$$\begin{bmatrix} 1 \\ -1 \\ 0 \\ 0 \end{bmatrix}$$

is in the null space of A. Form
$$B' = \begin{bmatrix} 1 & 1 & 0 & 3 \\ 0 & 0 & 1 & 2 \\ 1 & -1 & 0 & 0 \end{bmatrix}.$$

Now
$$RREF(B') = \begin{bmatrix} 1 & 0 & 0 & \frac{3}{2} \\ 0 & 1 & 0 & \frac{3}{2} \\ 0 & 0 & 1 & 2 \end{bmatrix},$$

so
$$H(B') = \begin{bmatrix} 1 & 0 & 0 & \frac{3}{2} \\ 0 & 1 & 0 & \frac{3}{2} \\ 0 & 0 & 1 & 2 \\ 0 & 0 & 0 & 0 \end{bmatrix},$$

and thus
$$\begin{bmatrix} \frac{3}{2} \\ \frac{3}{2} \\ 2 \\ -1 \end{bmatrix}$$

is in the null space of B' and so is
$$\begin{bmatrix} 3 \\ 3 \\ 4 \\ -2 \end{bmatrix}.$$

Now we are done and
$$\{(1, -1, 0, 0), (3, 3, 4, -2)\}$$

is an orthogonal basis of W.

Fact *If the rows of B are a basis of the null space of C, then null space of B = row space of C.*

Next we find an orthogonal basis for a subspace S when S described by a basis or a spanning set,

$$S = sp\{\mathbf{u}_1, \mathbf{u}_2, \cdots \mathbf{u}_j\}.$$

Procedure (2) *Put $\mathbf{u}_1, \mathbf{u}_2, \cdots \mathbf{u}_j$ as the rows of a matrix C, so that S equals the row space of C. Find a basis $\{\mathbf{v}_1, \mathbf{v}_2, \cdots, \mathbf{v}_k\}$ of the null space of C. (If the null space of C is only the zero vector, then S is all of R^n and you may take the standard basis of R^n as your orthogonal basis of S.) Let*

$$B = \begin{bmatrix} \mathbf{v}_1 \\ \mathbf{v}_2 \\ \vdots \\ \mathbf{v}_k \end{bmatrix},$$

then S equals the null space of B. Now that S is described as the null space of a matrix, we can proceed exactly as in Procedure 1.

Example Find an orthogonal basis of S = the row space of A, where

$$A = \begin{bmatrix} 2 & 2 & 2 & 10 \\ 1 & 1 & 1 & 5 \\ 3 & 3 & -1 & 7 \\ 4 & 4 & 3 & 18 \end{bmatrix},$$

as in the preceding example. Again

$$H(A) = \begin{bmatrix} 1 & 1 & 0 & 3 \\ 0 & 0 & 0 & 0 \\ 0 & 0 & 1 & 2 \\ 0 & 0 & 0 & 0 \end{bmatrix}.$$

Now

$$\begin{bmatrix} 1 \\ -1 \\ 0 \\ 0 \end{bmatrix} \text{ and } \begin{bmatrix} 3 \\ 0 \\ 2 \\ -1 \end{bmatrix}$$

Diagonalization

form a basis of the null space of A, thus we let

$$B = \begin{bmatrix} 1 & -1 & 0 & 0 \\ 3 & 0 & 2 & -1 \end{bmatrix}.$$

Now

$$H(B) = \begin{bmatrix} 1 & 0 & \frac{2}{3} & -\frac{1}{3} \\ 0 & 1 & \frac{2}{3} & -\frac{1}{3} \\ 0 & 0 & 0 & 0 \\ 0 & 0 & 0 & 0 \end{bmatrix}.$$

If you look at B, it is clear by inspection that

$$\begin{bmatrix} 0 \\ 0 \\ 1 \\ 2 \end{bmatrix}$$

is in the null space of B. (You could also take a nonzero column of $(H(B) - I)$.) Thus we form

$$B' = \begin{bmatrix} 1 & 0 & \frac{2}{3} & -\frac{1}{3} \\ 0 & 1 & \frac{2}{3} & -\frac{1}{3} \\ 0 & 0 & 1 & 2 \end{bmatrix}.$$

Now

$$RREF(B') = \begin{bmatrix} 1 & 0 & 0 & -\frac{5}{3} \\ 0 & 1 & 0 & -\frac{5}{3} \\ 0 & 0 & 1 & 2 \end{bmatrix},$$

and

$$H(B') = \begin{bmatrix} 1 & 0 & 0 & -\frac{5}{3} \\ 0 & 1 & 0 & -\frac{5}{3} \\ 0 & 0 & 1 & 2 \\ 0 & 0 & 0 & 0 \end{bmatrix},$$

so

$$\begin{bmatrix} -\frac{5}{3} \\ -\frac{5}{3} \\ 2 \\ -1 \end{bmatrix}$$

is a vector in the null space of B'. If we prefer we can use

$$\begin{bmatrix} 5 \\ 5 \\ -6 \\ 3 \end{bmatrix}.$$

Thus
$$\{(0,0,1,2), (5,5,-6,3)\}$$
is an orthogonal basis of S, the row space of A.

Fact *If you combine an orthogonal basis of the row space of A with an orthogonal basis of the null space of A you get an orthogonal basis of R^n.*

If W is a subspace of R^n, the set of vectors in R^n orthogonal to all vectors in W is also a subspace of R^n called the **orthogonal complement** of W, or $W^\perp$.

Fact *If W is the row space of A, then the null space of A is $W^\perp$.*

The **length** of a vector $\mathbf{u}$ which we denote by $|\mathbf{u}|$ is $\sqrt{\mathbf{u} \cdot \mathbf{u}}$. If $\mathbf{u}$ is in R^n let $\mathbf{u} = (a_1, a_2, \cdots, a_n)$, then

$$|\mathbf{u}| = \sqrt{\mathbf{u} \cdot \mathbf{u}} = \sqrt{a_1^2 + a_2^2 + \cdots + a_n^2}.$$

An **orthonormal basis** of a nonzero subspace W of R^n is an orthogonal basis $\{\mathbf{v}_1, \mathbf{v}_2, \cdots, \mathbf{v}_k\}$ of W which also satisfies $\mathbf{v}_i \cdot \mathbf{v}_i = 1$ for all $i = 1, 2, \cdots, k$. One obtains an orthonormal basis from an orthogonal basis easily. If $\{\mathbf{u}_1, \mathbf{u}_2, \cdots, \mathbf{u}_k\}$ is an orthogonal basis, let

$$\mathbf{v}_i = \left(\frac{1}{\sqrt{\mathbf{u}_i \cdot \mathbf{u}_i}} \right) \mathbf{u}_i.$$

Then $\{\mathbf{v}_1, \mathbf{v}_2, \cdots, \mathbf{v}_k\}$ is an orthonormal basis. This process is called normalizing the basis vectors. We divide each vector by its length, thus obtaining vectors of length one, which are called **unit vectors**.

Diagonalization

Exercises

1. For each of the following matrices, find an orthogonal basis of the null space.

    (a) $\begin{bmatrix} 1 & 2 \\ 3 & 6 \end{bmatrix}$

    (b) $\begin{bmatrix} -2 & 1 & 1 \\ 1 & -2 & 1 \\ 1 & 1 & -2 \end{bmatrix}$

    (c) $\begin{bmatrix} 1 & 2 & 3 \\ 2 & 3 & 4 \\ -3 & -2 & -1 \end{bmatrix}$

    (d) $\begin{bmatrix} 1 & -4 & 0 & 1 \\ 3 & 3 & 0 & -2 \\ -2 & -1 & 0 & 1 \\ 8 & -5 & 3 & -1 \end{bmatrix}$

    (e) $\begin{bmatrix} 2 & 4 & 1 & 11 \\ 3 & 6 & 2 & 18 \\ 1 & 2 & 1 & 7 \\ 4 & 8 & -5 & 1 \end{bmatrix}$.

2. For each of the matrices in problem 1 determine an orthogonal basis of the row space.

3. Find orthonormal bases for the null spaces of the matrices in problem 1.

Section 3.4 Orthogonal Diagonalization

Recall that the transpose of a matrix A is the matrix A^T whose rows are the columns of A, or equivalently $[A^T]_{ij} = a_{ji}$, the ij entry in A^T is the ji entry in A.

Example *If*
$$A = \begin{bmatrix} 1 & 2 \\ 3 & 4 \\ 5 & 6 \end{bmatrix},$$
then
$$A^T = \begin{bmatrix} 1 & 3 & 5 \\ 2 & 4 & 6 \end{bmatrix}.$$

A square matrix A is **symmetric** if $A^T = A$.

Example
$$\begin{bmatrix} 1 & 2 & 3 \\ 2 & 0 & 4 \\ 3 & 4 & 7 \end{bmatrix}$$
is a symmetric matrix.

A square matrix P is **orthogonal** if $P^T = P^{-1}$.

Example
$$\begin{bmatrix} \frac{1}{\sqrt{2}} & \frac{1}{\sqrt{2}} \\ -\frac{1}{\sqrt{2}} & \frac{1}{\sqrt{2}} \end{bmatrix}$$
is an orthogonal matrix since
$$\begin{bmatrix} \frac{1}{\sqrt{2}} & \frac{1}{\sqrt{2}} \\ -\frac{1}{\sqrt{2}} & \frac{1}{\sqrt{2}} \end{bmatrix} \begin{bmatrix} \frac{1}{\sqrt{2}} & -\frac{1}{\sqrt{2}} \\ \frac{1}{\sqrt{2}} & \frac{1}{\sqrt{2}} \end{bmatrix} = \begin{bmatrix} 1 & 0 \\ 0 & 1 \end{bmatrix}.$$

A matrix A is **orthogonally diagonalizable** if there exists an orthogonal matrix P such that $P^{-1}AP$ equals a diagonal matrix D.

Fact *An $n \times n$ matrix P is orthogonal if and only if its columns (or rows) form an orthonormal basis of R^n.*

Reason. The ij entry in $P^T P$ is the dot product of column i of P and column j of P.

Warning *For P to be orthogonal, its columns (or rows) must be orthonormal, not just orthogonal.*

Diagonalization

Fact *If A is a (real) symmetric matrix, then A is orthogonally diagonalizable. Moreover to be orthogonally diagonalizable a matrix must be symmetric.*

Procedure *Recall that to find a P such that $P^{-1}AP$ is diagonal, we first find the eigenvalues of A, and then a basis for each eigenspace, then these basis vectors become the columns of P. To make P orthogonal, we want to choose an orthonormal basis for each eigenspace. (For each eigenvalue r of A, find an orthonormal basis of the null space of $(A - rI)$ as explained in the previous section.) Put these basis vectors as the columns of P, then P is orthogonal and $P^{-1}AP$ is a diagonal matrix.*

Fact *If A is a symmetric matrix, eigenvectors of A going with different eigenvalues are orthogonal.*

Fact *If A is a symmetric matrix, the eigenvalues of A are all real and the geometric multiplicity equals the algebraic multiplicity for every eigenvalue.*

Example Find a matrix P which orthogonally diagonalizes A when

$$A = \begin{bmatrix} 4 & 1 \\ 1 & 4 \end{bmatrix}.$$

Solution:

$$\begin{vmatrix} 4-k & 1 \\ 1 & 4-k \end{vmatrix} = k^2 - 8k + 15 = (k-3)(k-5).$$

So the eigenvalues are 3 and 5. For $k = 3$ we have

$$A - 3I = \begin{bmatrix} 1 & 1 \\ 1 & 1 \end{bmatrix},$$

and

$$\begin{bmatrix} 1 \\ -1 \end{bmatrix}$$

is a basis of the null space. For $k = 5$ we have

$$A - 5I = \begin{bmatrix} -1 & 1 \\ 1 & -1 \end{bmatrix},$$

and
$$\begin{bmatrix} 1 \\ 1 \end{bmatrix}$$
is a basis of the null space. Note that these two vectors are orthogonal, so
$$\begin{bmatrix} 1 \\ -1 \end{bmatrix} \text{ and } \begin{bmatrix} 1 \\ 1 \end{bmatrix}$$
form an orthogonal basis of R^2. For the columns of P we need an orthonormal basis of R^2, so we need to normalize. The length of each of these vectors is $\sqrt{2}$, so we divide each vector by $\sqrt{2}$ to get vectors of length 1. Thus
$$P = \begin{bmatrix} \frac{1}{\sqrt{2}} & \frac{1}{\sqrt{2}} \\ -\frac{1}{\sqrt{2}} & \frac{1}{\sqrt{2}} \end{bmatrix}$$
is an orthogonal matrix which diagonalizes A. You can verify that $AP = PD$, where
$$D = \begin{bmatrix} 3 & 0 \\ 0 & 5 \end{bmatrix}.$$

Example Find a matrix P which orthogonally diagonalizes
$$A = \begin{bmatrix} 1 & 2 & 1 \\ 2 & 4 & 2 \\ 1 & 2 & 1 \end{bmatrix}.$$

Solution:
$$\begin{vmatrix} 1-k & 2 & 1 \\ 2 & 4-k & 2 \\ 1 & 2 & 1-k \end{vmatrix} = -k^3 + 6k^2 = -k^2(k-6).$$

The eigenvalues are 0 and 6. For $k = 6$ we have
$$A - 6I = \begin{bmatrix} -5 & 2 & 1 \\ 2 & -2 & 2 \\ 1 & 2 & -5 \end{bmatrix},$$

Diagonalization

which has $\begin{bmatrix} 1 \\ 2 \\ 1 \end{bmatrix}$ in its null space. We normalize to obtain

$$\begin{bmatrix} \frac{1}{\sqrt{6}} \\ \frac{2}{\sqrt{6}} \\ \frac{1}{\sqrt{6}} \end{bmatrix}.$$

For the null space of

$$A - 0I = \begin{bmatrix} 1 & 2 & 1 \\ 2 & 4 & 2 \\ 1 & 2 & 1 \end{bmatrix},$$

observe that it is also the null space of

$$\begin{bmatrix} 1 & 2 & 1 \end{bmatrix}.$$

By inspection we see that $\begin{bmatrix} 1 \\ -1 \\ 1 \end{bmatrix}$ is in that null space. We form

$$\begin{bmatrix} 1 & 2 & 1 \\ 1 & -1 & 1 \end{bmatrix}.$$

The Hermite form of this matrix is

$$\begin{bmatrix} 1 & 0 & 1 \\ 0 & 1 & 0 \\ 0 & 0 & 0 \end{bmatrix},$$

thus $\begin{bmatrix} 1 \\ 0 \\ -1 \end{bmatrix}$ is the second eigenvector we want for $k = 0$.

$$\begin{bmatrix} 1 \\ -1 \\ 1 \end{bmatrix} \text{ and } \begin{bmatrix} 1 \\ 0 \\ -1 \end{bmatrix}$$

form an orthogonal basis for the eigenspace going with $k = 0$. Normalizing gives

$$\begin{bmatrix} \frac{1}{\sqrt{3}} \\ -\frac{1}{\sqrt{3}} \\ \frac{1}{\sqrt{3}} \end{bmatrix} \text{ and } \begin{bmatrix} \frac{1}{\sqrt{2}} \\ 0 \\ -\frac{1}{\sqrt{2}} \end{bmatrix}$$

as an orthonormal basis of this eigenspace. Thus

$$P = \begin{bmatrix} \frac{1}{\sqrt{6}} & \frac{1}{\sqrt{3}} & \frac{1}{\sqrt{2}} \\ \frac{2}{\sqrt{6}} & -\frac{1}{\sqrt{3}} & 0 \\ \frac{1}{\sqrt{6}} & \frac{1}{\sqrt{3}} & -\frac{1}{\sqrt{2}} \end{bmatrix}$$

will orthogonaly diagonalize A. $P^{-1}AP$ will equal

$$D = \begin{bmatrix} 6 & 0 & 0 \\ 0 & 0 & 0 \\ 0 & 0 & 0 \end{bmatrix}.$$

Check:

$$AP = \begin{bmatrix} 1 & 2 & 1 \\ 2 & 4 & 2 \\ 1 & 2 & 1 \end{bmatrix} \begin{bmatrix} \frac{1}{\sqrt{6}} & \frac{1}{\sqrt{3}} & \frac{1}{\sqrt{2}} \\ \frac{2}{\sqrt{6}} & -\frac{1}{\sqrt{3}} & 0 \\ \frac{1}{\sqrt{6}} & \frac{1}{\sqrt{3}} & -\frac{1}{\sqrt{2}} \end{bmatrix} = \begin{bmatrix} \frac{6}{\sqrt{6}} & 0 & 0 \\ \frac{12}{\sqrt{6}} & 0 & 0 \\ \frac{6}{\sqrt{6}} & 0 & 0 \end{bmatrix} = PD.$$

Example Find a matrix which orthogonally diagonalizes

$$A = \begin{bmatrix} 2 & 2 & 1 & 0 \\ 2 & -1 & -2 & 0 \\ 1 & -2 & 2 & 0 \\ 0 & 0 & 0 & 3 \end{bmatrix}.$$

Solution:

$$\begin{vmatrix} 2-k & 2 & 1 & 0 \\ 2 & -1-k & -2 & 0 \\ 1 & -2 & 2-k & 0 \\ 0 & 0 & 0 & 3-k \end{vmatrix} = (3-k) \begin{vmatrix} 2-k & 2 & 1 \\ 2 & -1-k & -2 \\ 1 & -2 & 2-k \end{vmatrix}$$

$$= (k-3)(-k^3 + 3k + 9k - 27)$$
$$= (k-3)(k^3 - 3k - 9k + 27)$$
$$= (k-3)(k-3)(k^2 - 9)$$
$$= (k-3)^3(k+3).$$

Diagonalization

Thus the eigenvalues are 3 and -3. For $k = -3$,

$$A + 3I = \begin{bmatrix} 5 & 2 & 1 & 0 \\ 2 & 2 & -2 & 0 \\ 1 & -2 & 5 & 0 \\ 0 & 0 & 0 & 6 \end{bmatrix},$$

which has $\begin{bmatrix} -1 \\ 2 \\ 1 \\ 0 \end{bmatrix}$ *in its null space. Normalizing we obtain*

$$\begin{bmatrix} -\frac{1}{\sqrt{6}} \\ \frac{2}{\sqrt{6}} \\ \frac{1}{\sqrt{6}} \\ 0 \end{bmatrix}.$$

For $k = 3$,

$$A - 3I = \begin{bmatrix} -1 & 2 & 1 & 0 \\ 2 & -4 & -2 & 0 \\ 1 & -2 & -1 & 0 \\ 0 & 0 & 0 & 0 \end{bmatrix}.$$

$(A - 3I)$ has the same null space as

$$\begin{bmatrix} -1 & 2 & 1 & 0 \end{bmatrix}.$$

By inspection $(1, 0, 1, 0)$ is in this null space, so we form

$$\begin{bmatrix} 1 & -2 & -1 & 0 \\ 1 & 0 & 1 & 0 \end{bmatrix},$$

and clearly $(0, 0, 0, 1)$ is in the null space of this matrix. Thus we form

$$B = \begin{bmatrix} 1 & -2 & -1 & 0 \\ 1 & 0 & 1 & 0 \\ 0 & 0 & 0 & 1 \end{bmatrix}.$$

Then

$$H(B) = \begin{bmatrix} 1 & 0 & 1 & 0 \\ 0 & 1 & 1 & 0 \\ 0 & 0 & 0 & 0 \\ 0 & 0 & 0 & 1 \end{bmatrix}.$$

So $\begin{bmatrix} 1 \\ 1 \\ -1 \\ 0 \end{bmatrix}$ is the third vector we want, and

$$\begin{bmatrix} 1 \\ 0 \\ 1 \\ 0 \end{bmatrix}, \begin{bmatrix} 0 \\ 0 \\ 0 \\ 1 \end{bmatrix}, \text{ and } \begin{bmatrix} 1 \\ 1 \\ -1 \\ 0 \end{bmatrix}$$

form an orthogonal basis of this eigenspace. Normalizing gives

$$\begin{bmatrix} \frac{1}{\sqrt{2}} \\ 0 \\ \frac{1}{\sqrt{2}} \\ 0 \end{bmatrix}, \begin{bmatrix} 0 \\ 0 \\ 0 \\ 1 \end{bmatrix}, \text{ and } \begin{bmatrix} \frac{1}{\sqrt{3}} \\ \frac{1}{\sqrt{3}} \\ -\frac{1}{\sqrt{3}} \\ 0 \end{bmatrix}$$

as an orthonormal basis of this eigenspace. Now

$$P = \begin{bmatrix} -\frac{1}{\sqrt{6}} & \frac{1}{\sqrt{2}} & 0 & \frac{1}{\sqrt{3}} \\ \frac{2}{\sqrt{6}} & 0 & 0 & \frac{1}{\sqrt{3}} \\ \frac{1}{\sqrt{6}} & \frac{1}{\sqrt{2}} & 0 & -\frac{1}{\sqrt{3}} \\ 0 & 0 & 1 & 0 \end{bmatrix}$$

is a matrix that will orthogonally diagonalize A. The reader may check that $PP^T = I$ and that

$$AP = P \begin{bmatrix} -3 & 0 & 0 & 0 \\ 0 & 3 & 0 & 0 \\ 0 & 0 & 3 & 0 \\ 0 & 0 & 0 & 3 \end{bmatrix}.$$

Exercises

1. For each of the following matrices A, find an orthogonal matrix P such that $P^{-1}AP = D$, where D is diagonal, and check by showing that $PP^T = I$ and $AP = PD$.

   (a) $\begin{bmatrix} 4 & 1 \\ 1 & 2 \end{bmatrix}$

Diagonalization

(b) $\begin{bmatrix} 3 & 2 \\ 2 & 3 \end{bmatrix}$

(c) $\begin{bmatrix} 5 & -2 & 2 \\ -2 & 2 & 4 \\ 2 & 4 & 2 \end{bmatrix}$

(d) $\begin{bmatrix} 4 & -2 & 0 \\ -2 & 3 & 2 \\ 0 & 2 & 2 \end{bmatrix}$

(e) $\begin{bmatrix} 4 & 1 & 1 \\ 1 & 4 & 1 \\ 1 & 1 & 4 \end{bmatrix}$

(f) $\begin{bmatrix} -3 & 1 & 1 & 1 \\ 1 & -3 & 1 & 1 \\ 1 & 1 & -3 & 1 \\ 1 & 1 & 1 & -3 \end{bmatrix}$

(g) $\begin{bmatrix} 2 & 1 & 0 & 0 \\ 1 & 2 & 0 & 0 \\ 0 & 0 & 4 & 3 \\ 0 & 0 & 3 & 4 \end{bmatrix}$.

2. If A and B are $n \times n$ matrices, show that $(AB)^T = B^T A^T$.

3. Use exercise 2 to show that if P is orthogonal, D diagonal and A an $n \times n$ matrix such that $P^{-1}AP = D$, then A must be symmetric.

Remark The following procedure is another way to obtain an orthogonal basis for a subspace.

(**Gram-Schmidt**)To find an orthogonal basis of a subspace S given a basis $\mathbf{u}_1, \mathbf{u}_2, \cdots, \mathbf{u}_n$ of S:

(a) Let
$$\mathbf{v}_1 = \mathbf{u}_1.$$

(b) Let
$$\mathbf{v}_2 = \mathbf{u}_2 - \left(\frac{\mathbf{u}_2 \cdot \mathbf{v}_1}{\mathbf{v}_1 \cdot \mathbf{v}_1}\right) \mathbf{v}_1.$$

(c) Let
$$\mathbf{v}_3 = \mathbf{u}_3 - \left(\frac{\mathbf{u}_3 \cdot \mathbf{v}_1}{\mathbf{v}_1 \cdot \mathbf{v}_1}\right)\mathbf{v}_1 - \left(\frac{\mathbf{u}_3 \cdot \mathbf{v}_2}{\mathbf{v}_2 \cdot \mathbf{v}_2}\right)\mathbf{v}_2.$$

(d) Let
$$\mathbf{v}_4 = \mathbf{u}_4 - \left(\frac{\mathbf{u}_4 \cdot \mathbf{v}_1}{\mathbf{v}_1 \cdot \mathbf{v}_1}\right)\mathbf{v}_1 - \left(\frac{\mathbf{u}_4 \cdot \mathbf{v}_2}{\mathbf{v}_2 \cdot \mathbf{v}_2}\right)\mathbf{v}_2 - \left(\frac{\mathbf{u}_4 \cdot \mathbf{v}_3}{\mathbf{v}_3 \cdot \mathbf{v}_3}\right)\mathbf{v}_3.$$

(e) Continue to
$$\mathbf{v}_n = \mathbf{u}_n - \left(\frac{\mathbf{u}_n \cdot \mathbf{v}_1}{\mathbf{v}_1 \cdot \mathbf{v}_1}\right)\mathbf{v}_1 - \left(\frac{\mathbf{u}_n \cdot \mathbf{v}_2}{\mathbf{v}_2 \cdot \mathbf{v}_2}\right)\mathbf{v}_2 \cdots - \left(\frac{\mathbf{u}_n \cdot \mathbf{v}_{n-1}}{\mathbf{v}_{n-1} \cdot \mathbf{v}_{n-1}}\right)\mathbf{v}_{n-1}.$$

(f) Now $\{\mathbf{v}_1, \mathbf{v}_2, \cdots, \mathbf{v}_n\}$ is an orthogonal basis of S.

The process given above amounts to subtracting from each $\mathbf{u}_i$ its projection on the preceding $\mathbf{v}_i$, starting off with $\mathbf{v}_1 = \mathbf{u}_1$. If you want an orthonormal basis, just divide each $\mathbf{v}_i$ by its length.

Chapter 4

Applications

Section 4.1 Linear Equations

As we have seen the matrix equation $A\mathbf{x} = \mathbf{b}$, where A is an $m \times n$ matrix, $\mathbf{x}$ is a column of n unknowns, and $\mathbf{b}$ is a column of m numbers, represents a system of m linear equations in n unknowns. We solve this system by finding the RREF of the augmented matrix $[A\ \mathbf{b}]$. This system has a solution if and only if $\mathbf{b}$ is in the column space of A, and in fact the solution tells you how to write the vector $\mathbf{b}$ as a linear combination of the columns of A.

Fact $A\mathbf{x} = \mathbf{b}$ *has a solution if and only if $\mathbf{b}$ is in the column space of A.*

Example *Solve the system of equations*

$$\begin{aligned} 3x + 2y - z &= 1 \\ x - y + 3z &= 2 \\ 2x + 3y - 4z &= 3 \end{aligned}$$

or equivalently

$$x \begin{bmatrix} 3 \\ 1 \\ 2 \end{bmatrix} + y \begin{bmatrix} 2 \\ -1 \\ 3 \end{bmatrix} + z \begin{bmatrix} -1 \\ 3 \\ -4 \end{bmatrix} = \begin{bmatrix} 1 \\ 2 \\ 3 \end{bmatrix}.$$

Solution: The augmented matrix of this system is:

$$[A \ b] = \begin{bmatrix} 3 & 2 & -1 & 1 \\ 1 & -1 & 3 & 2 \\ 2 & 3 & -4 & 3 \end{bmatrix},$$

and

$$RREF[A \ b] = \begin{bmatrix} 1 & 0 & 1 & 0 \\ 0 & 1 & -2 & 0 \\ 0 & 0 & 0 & 1 \end{bmatrix}.$$

The last column of $RREF[A \ b]$ is not a linear combination of the first three, thus the last column of $[A \ b]$ is not a linear combination of the columns of A, and so the system has no solution.

Fact $Ax = b$ has a solution if and only if $rank(A) = rank[A \ b]$.

Fact $Ax = b$ has a solution if and only if $RREF[A \ b]$ has no leading one in the last column.

Fact If $b \neq O$, the solution set of $Ax = b$ is not a subspace of R^n.

Reason. The sum of two solutions is not a solution, since if $Au = b$ and $Av = b$, then $A(u + v) = Au + Av = b + b = 2b \neq b$.

The system $Ax = O$ is called **homogeneous** ($b = O$). The solution set of $Ax = O$ is the null space of A. If a solution to $Ax = b$ exists for a given value of b, the solution set of this system is related to the null space of A in the following way:

Fact If v is one solution to $Ax = b$, then the solution set of $Ax = b$ is all vectors of the form $v + w$, where w is in the null space of A.

Reason. Clearly if $Av = b$ and $Aw = O$, then $A(v+w) = Av+Aw = b+O = b$, so all vectors of this form are solutions of $Ax = b$. Moreover if u is any other solution to $Ax = b$, then $A(u - v) = Au - Av = b - b = O$, so $(u - v)$ is in the null space of A, and thus $(u - v) = w$ for some w in the null space of A, thus $u = v + w$, where w is in the null space of A.

Applications

Example *Solve the system of equations*

$$\begin{aligned} 5x_1 + 2x_2 + 4x_3 + 31x_4 &= -2 \\ 2x_1 + x_2 + x_3 + 13x_4 &= 0 \\ -x_1 + x_2 - 5x_3 - 2x_4 &= 6 \\ 4x_1 + 8x_3 + 20x_4 &= -8 \end{aligned}.$$

Solution: Let

$$A = \begin{bmatrix} 5 & 2 & 4 & 31 \\ 2 & 1 & 1 & 13 \\ -1 & 1 & -5 & -2 \\ 4 & 0 & 8 & 20 \end{bmatrix}$$

and

$$\mathbf{b} = \begin{bmatrix} -2 \\ 0 \\ 6 \\ -8 \end{bmatrix}.$$

The augmented matrix of the system $A\mathbf{x} = \mathbf{b}$ *is*

$$[\,A\ \mathbf{b}\,] = \begin{bmatrix} 5 & 2 & 4 & 31 & -2 \\ 2 & 1 & 1 & 13 & 0 \\ -1 & 1 & -5 & -2 & 6 \\ 4 & 0 & 8 & 20 & -8 \end{bmatrix},$$

and

$$RREF\,[\,A\ \mathbf{b}\,] = \begin{bmatrix} 1 & 0 & 2 & 5 & -2 \\ 0 & 1 & -3 & 3 & 4 \\ 0 & 0 & 0 & 0 & 0 \\ 0 & 0 & 0 & 0 & 0 \end{bmatrix}.$$

Looking at $RREF\,[\,A\ \mathbf{b}\,]$, *we see that the last column is* -2 *times the first plus* 4 *times the second. This same relation holds between the columns of* $[\,A\ \mathbf{b}\,]$, *and therefore* $\begin{bmatrix} -2 \\ 4 \\ 0 \\ 0 \end{bmatrix}$ *is a solution of* $A\mathbf{x} = \mathbf{b}$.
To find the rest of the solutions, we need only find the null space of A.

We have
$$RREF(A) = \begin{bmatrix} 1 & 0 & 2 & 5 \\ 0 & 1 & -3 & 3 \\ 0 & 0 & 0 & 0 \\ 0 & 0 & 0 & 0 \end{bmatrix} = H(A),$$

and
$$(H(A) - I) = \begin{bmatrix} 0 & 0 & 2 & 5 \\ 0 & 0 & -3 & 3 \\ 0 & 0 & -1 & 0 \\ 0 & 0 & 0 & -1 \end{bmatrix},$$

so that $\begin{bmatrix} 2 \\ -3 \\ -1 \\ 0 \end{bmatrix}$ and $\begin{bmatrix} 5 \\ 3 \\ 0 \\ -1 \end{bmatrix}$ form a basis for the null space of A. The vectors in the null space of A are thus of the form

$$r \begin{bmatrix} 2 \\ -3 \\ -1 \\ 0 \end{bmatrix} + s \begin{bmatrix} 5 \\ 3 \\ 0 \\ -1 \end{bmatrix},$$

where r and s are any numbers. Thus the solutions of $A\mathbf{x} = \mathbf{b}$ are all vectors of the form

$$\mathbf{x} = \begin{bmatrix} -2 \\ 4 \\ 0 \\ 0 \end{bmatrix} + r \begin{bmatrix} 2 \\ -3 \\ -1 \\ 0 \end{bmatrix} + s \begin{bmatrix} 5 \\ 3 \\ 0 \\ -1 \end{bmatrix}.$$

In terms of the unknowns x_1, x_2, x_3, x_4 this may be written
$$\begin{aligned} x_1 &= -2 + 2r + 5s \\ x_2 &= 4 - 3r + 3s \\ x_3 &= -r \\ x_4 &= -s \end{aligned}$$

Example Solve the system of equations
$$\begin{aligned} 3x + 2y - z &= 1 \\ x - y + 3z &= 2 \\ 2x + 3y - 4z &= -1 \\ 2x + 3y + 3z &= 4 \end{aligned}.$$

Applications

Solution: The augmented matrix of this system is

$$[A \ b] = \begin{bmatrix} 3 & 2 & -1 & 1 \\ 1 & -1 & 3 & 2 \\ 2 & 3 & -4 & -1 \\ 2 & 3 & 3 & 4 \end{bmatrix},$$

and

$$RREF[A \ b] = \begin{bmatrix} 1 & 0 & 0 & \frac{2}{7} \\ 0 & 1 & 0 & \frac{3}{7} \\ 0 & 0 & 1 & \frac{5}{7} \\ 0 & 0 & 0 & 0 \end{bmatrix}.$$

From the RREF we see that $x = \frac{2}{7}$, $y = \frac{3}{7}$, and $z = \frac{5}{7}$ is a solution. Since the null space of A contains only the zero vector, this is the only solution.

Example Solve the system of equations

$$\begin{aligned} 3x + 2y - z &= 1 \\ x - y + 3z &= 2 \\ 2x + 3y - 4z &= -1 \end{aligned}.$$

Solution: The augmented matrix is

$$[A \ b] = \begin{bmatrix} 3 & 2 & -1 & 1 \\ 1 & -1 & 3 & 2 \\ 2 & 3 & -4 & -1 \end{bmatrix},$$

and

$$RREF[A \ b] = \begin{bmatrix} 1 & 0 & 1 & 1 \\ 0 & 1 & -2 & -1 \\ 0 & 0 & 0 & 0 \end{bmatrix}.$$

From $RREF[A \ b]$ we see that the last column is 1 times the first column plus -1 times the second column. Thus $x = 1$, $y = -1$, $z = 0$, or

$$\begin{bmatrix} x \\ y \\ z \end{bmatrix} = \begin{bmatrix} 1 \\ -1 \\ 0 \end{bmatrix}$$

is a solution. It is not the only solution, because nullspace$(A) \neq \mathbf{O}$.

$$RREF(A) = \begin{bmatrix} 1 & 0 & 1 \\ 0 & 1 & -2 \\ 0 & 0 & 0 \end{bmatrix} = H(A),$$

and

$$(H(A) - I) = \begin{bmatrix} 0 & 0 & 1 \\ 0 & 0 & -2 \\ 0 & 0 & -1 \end{bmatrix},$$

so a basis for nullspace of A is

$$\begin{bmatrix} 1 \\ -2 \\ -1 \end{bmatrix}.$$

Thus the complete solution set for our original system of equations is all vectors of the form

$$\begin{bmatrix} x \\ y \\ z \end{bmatrix} = \begin{bmatrix} 1 \\ -1 \\ 0 \end{bmatrix} + r \begin{bmatrix} 1 \\ -2 \\ -1 \end{bmatrix}.$$

Example In an electrical circuit involving two power sources of 12 volts and three resistors of 3, 5, and 7 ohms the currents $i_1, i_2,$ and i_3 satisfy the following equations according to Kirchoff's Law:

$$\begin{array}{rcrcrcr} i_1 & - & i_2 & - & i_3 & = & 0 \\ 5i_1 & + & 7i_2 & & & = & 24 \\ 5i_1 & & & + & 3i_3 & = & 12 \\ & & -7i_2 & + & 3i_3 & = & -12 \end{array}.$$

Solve for $i_1, i_2,$ and i_3 if possible. Solution: The augmented matrix of the system is

$$\begin{bmatrix} A & \mathbf{b} \end{bmatrix} = \begin{bmatrix} 1 & -1 & -1 & 0 \\ 5 & 7 & 0 & 24 \\ 5 & 0 & 3 & 12 \\ 0 & -7 & 3 & -12 \end{bmatrix},$$

and

$$RREF\begin{bmatrix} A & b \end{bmatrix} = \begin{bmatrix} 1 & 0 & 0 & \frac{156}{71} \\ 0 & 1 & 0 & \frac{132}{71} \\ 0 & 0 & 1 & \frac{24}{71} \\ 0 & 0 & 0 & 0 \end{bmatrix}.$$

We see that the system has a unique solution

$$i_1 = \frac{156}{71}, \quad i_2 = \frac{132}{71}, \quad i_3 = \frac{24}{71}.$$

Example *Suppose a nonhomogeneous system of equations $A\mathbf{x} = \mathbf{b}$ has as the RREF of its augmented matrix*

$$RREF\begin{bmatrix} A & b \end{bmatrix} = \begin{bmatrix} 1 & 0 & 2 & 0 & 8 \\ 0 & 1 & 3 & 0 & 11 \\ 0 & 0 & 0 & 1 & 2 \\ 0 & 0 & 0 & 0 & 0 \end{bmatrix}.$$

Since the last column is 8 times the first plus 11 times the second plus 2 times the fourth, this will also be true of the matrix $\begin{bmatrix} A & b \end{bmatrix}$, so one solution of $A\mathbf{x} = \mathbf{b}$ is

$$\mathbf{x} = \begin{bmatrix} 8 \\ 11 \\ 0 \\ 2 \end{bmatrix}$$

. (Note this is not the last column of $RREF\begin{bmatrix} A & b \end{bmatrix}$, but if we rearrange the rows so that the leading ones are on the diagonal, this solution vector will appear in the last column).
A basis for $nullspace(A)$ is

$$\begin{bmatrix} 2 \\ 3 \\ -1 \\ 0 \end{bmatrix},$$

so the general solution is

$$\begin{bmatrix} 8 + 2r \\ 11 + 3r \\ -r \\ 2 \end{bmatrix}.$$

Remark One can also get the solutions of the previous example by looking at the equations represented by

$$RREF\begin{bmatrix} A & b \end{bmatrix} = \begin{bmatrix} 1 & 0 & 2 & 0 & 8 \\ 0 & 1 & 3 & 0 & 11 \\ 0 & 0 & 0 & 1 & 2 \\ 0 & 0 & 0 & 0 & 0 \end{bmatrix},$$

which are (working from the bottom up)

$$\begin{aligned} x_4 &= 2 \\ x_2 &= 11 - 3x_3 \\ x_1 &= 8 - 2x_3 \end{aligned}$$

Letting $x_3 = t$, we get

$$\begin{bmatrix} x_1 \\ x_2 \\ x_3 \\ x_4 \end{bmatrix} = \begin{bmatrix} 8 - 2t \\ 11 - 3t \\ t \\ 2 \end{bmatrix},$$

where t can be any number. This agrees with our previous solution if we let $t = -r$. The second method above shows that when a solution of $A\mathbf{x} = \mathbf{b}$ exists, the variables corresponding to columns of $RREF\begin{bmatrix} A & b \end{bmatrix}$ which have a leading one can be solved for in terms of the remaining variables.

Exercises

1. Solve the system
$$\begin{aligned} 3x + 2y - z &= 1 \\ -3y + 3z &= 2 \\ 2z &= 8 \end{aligned}.$$

2. Determine whether $(3, 2, 5)$ is a linear combination of $(1, 3, 0)$, $(2, 1, 4)$, and $(4, 7, 4)$.

3. Find all solutions of the following system of linear equations.
$$\begin{aligned} 2w + x + 5y + z &= 0 \\ w - 3x + y + z &= 0 \\ 2w + y - z &= 0 \end{aligned}.$$

Applications

4. Find all solutions of the following system of linear equations.
$$\begin{aligned} u + 2v + 5w + 2x + 4y &= 0 \\ v + 3w\phantom{{}+2x} + 5y &= 0 \\ u + v + 2w + 4x + 9y &= 0 \end{aligned}$$

5. Find all solutions of the system
$$\begin{aligned} u + 2v + 5w + 2x + 4y &= 1 \\ v + 3w\phantom{{}+2x} + 5y &= 2 \\ u + v + 2w + 4x + 9y &= 2 \end{aligned}$$

Section 4.2 Differential Equations

A system of first order linear differential equations is a system relating n differentiable functions to their first derivatives. It is assumed that the derivative of each function is a linear combination of all the functions as follows: ($'$ indicates differentiation with respect to t).

$$\begin{aligned} x_1'(t) &= a_{11}x_1(t) + a_{12}x_2(t) + \cdots + a_{1n}x_n(t) \\ x_2'(t) &= a_{21}x_1(t) + a_{22}x_2(t) + \cdots + a_{2n}x_n(t) \\ &\vdots \\ x_n'(t) &= a_{n1}x_1(t) + a_{n2}x_2(t) + \cdots + a_{nn}x_n(t), \end{aligned}$$

which can be written in matrix form as

$$\begin{bmatrix} x_1'(t) \\ x_2'(t) \\ \vdots \\ x_n'(t) \end{bmatrix} = A \begin{bmatrix} x_1(t) \\ x_2(t) \\ \vdots \\ x_n(t) \end{bmatrix},$$

or even more simply as
$$\mathbf{x}'(t) = A\mathbf{x}(t),$$

where $\mathbf{x}(t)$ is the vector whose components are $x_i(t)$.

For $n = 1$ the system becomes
$$x_1'(t) = a_{11}x_1(t),$$

so we drop the subscripts and just write
$$x'(t) = ax(t).$$
In this case the variables separate if we rewrite the equation as
$$\frac{x'(t)}{x(t)} = a.$$
We can integrate both sides with respect to t and get
$$\ln(x) = at + b,$$
or
$$x(t) = e^{at+b} = e^{at}e^b = e^b e^{at}.$$
Substituting $t = 0$ we get $x(0) = e^b$. Thus we have that a solution is
$$x(t) = x(0)e^{at}.$$

If the coefficient matrix A of our system is diagonal, then the system is just n separate equations of this type
$$\begin{aligned} x_1'(t) &= a_{11}x_1(t) \\ x_2'(t) &= a_{22}x_2(t) \\ &\vdots \\ x_n'(t) &= a_{nn}x_n(t) \end{aligned}$$
so a solution would be
$$\begin{aligned} x_1(t) &= k_1 e^{a_{11}t} \\ x_2(t) &= k_2 e^{a_{22}t} \\ &\vdots \\ x_n(t) &= k_n e^{a_{nn}t} \end{aligned}$$
where the k_i are constants, in this case $k_i = x_i(0)$.

Now if A is similar to a diagonal matrix D, we can reduce the case $\mathbf{x}'(t) = A\mathbf{x}(t)$ to the diagonal case by a change of variable. Let P be an invertible matrix such that
$$P^{-1}AP = D = \begin{bmatrix} d_1 & 0 & \cdots & 0 \\ 0 & d_2 & & \vdots \\ \vdots & & \ddots & 0 \\ 0 & \cdots & 0 & d_n \end{bmatrix},$$

Applications

where the d_i are the eigenvalues of A. Then make the substitution

$$\mathbf{x}(t) = P\mathbf{u}(t),$$

or

$$\begin{bmatrix} x_1(t) \\ x_2(t) \\ \vdots \\ x_n(t) \end{bmatrix} = P \begin{bmatrix} u_1(t) \\ u_2(t) \\ \vdots \\ u_n(t) \end{bmatrix}.$$

Since the derivative of a linear combination of functions is a linear combination of the derivatives with the same coefficients, differentiating both sides gives

$$\begin{bmatrix} x_1'(t) \\ x_2'(t) \\ \vdots \\ x_n'(t) \end{bmatrix} = P \begin{bmatrix} u_1'(t) \\ u_2'(t) \\ \vdots \\ u_n'(t) \end{bmatrix}.$$

Substituting this in $\mathbf{x}'(t) = A\mathbf{x}(t)$ gives

$$P \begin{bmatrix} u_1'(t) \\ u_2'(t) \\ \vdots \\ u_n'(t) \end{bmatrix} = AP \begin{bmatrix} u_1(t) \\ u_2(t) \\ \vdots \\ u_n(t) \end{bmatrix},$$

or

$$\begin{bmatrix} u_1'(t) \\ u_2'(t) \\ \vdots \\ u_n'(t) \end{bmatrix} = P^{-1}AP \begin{bmatrix} u_1(t) \\ u_2(t) \\ \vdots \\ u_n(t) \end{bmatrix} = D \begin{bmatrix} u_1(t) \\ u_2(t) \\ \vdots \\ u_n(t) \end{bmatrix}.$$

This is now the diagonal case and we know the solution is

$$\begin{bmatrix} u_1(t) \\ u_2(t) \\ \vdots \\ u_n(t) \end{bmatrix} = \begin{bmatrix} k_1 e^{d_1 t} \\ k_2 e^{d_2 t} \\ \vdots \\ k_n e^{d_n t} \end{bmatrix}.$$

We can now find $\mathbf{x}(t)$ because $\mathbf{x}(t) = P\mathbf{u}(t)$.

Procedure *To solve the first order linear system of differential equations given by*

$$\mathbf{x}'(t) = A\mathbf{x}(t),$$

find the eigenvalues of A. If A is diagonalizable, find a matrix P such that $P^{-1}AP$ is diagonal. Then a solution is given by

$$\begin{bmatrix} x_1(t) \\ x_2(t) \\ \vdots \\ x_n(t) \end{bmatrix} = P \begin{bmatrix} k_1 e^{d_1 t} \\ k_2 e^{d_2 t} \\ \vdots \\ k_n e^{d_n t} \end{bmatrix},$$

where d_i are the eigenvalues of A in the same order as the corresponding eigenvectors appear as columns of P, and k_i are constants that we can find if we are given some initial conditions. To use this method it is essential that A be diagonalizable and that you be able to find both the eigenvalues of A (the d_i) and a matrix P that diagonalizes A.

Example *Solve the system of differential equations*

$$\begin{aligned} x_1' &= -1x_1 + 2x_2 \\ x_2' &= x_1 + 2x_2 + x_3 \\ x_3' &= 2x_2 - x_3 \end{aligned}.$$

Solution: Let

$$A = \begin{bmatrix} -1 & 2 & 0 \\ 1 & 2 & 1 \\ 0 & 2 & -1 \end{bmatrix}.$$

Then

$$|A - kI| = \begin{vmatrix} -1-k & 2 & 0 \\ 1 & 2-k & 1 \\ 0 & 2 & -1-k \end{vmatrix} = -k^3 + 7k + 6.$$

Solving

$$0 = k^3 - 7k - 6 = (k+1)(k+2)(k-3),$$

we see that the eigenvalues are -1, -2, and 3. Since they are all distinct, we know that A is diagonalizable. Using the methods of the previous chapter to find P (columns consisting of eigenvectors for -1, -2, and 3 respectively), we find that for

$$P = \begin{bmatrix} 1 & 2 & 1 \\ 0 & -1 & 2 \\ -1 & 2 & 1 \end{bmatrix}$$

and

$$D = \begin{bmatrix} -1 & 0 & 0 \\ 0 & -2 & 0 \\ 0 & 0 & 3 \end{bmatrix},$$

we have

$$P^{-1}AP = D.$$

Thus

$$\mathbf{x}(t) = P \begin{bmatrix} k_1 e^{-1t} \\ k_2 e^{-2t} \\ k_3 e^{3t} \end{bmatrix} = \begin{bmatrix} 1 & 2 & 1 \\ 0 & -1 & 2 \\ -1 & 2 & 1 \end{bmatrix} \begin{bmatrix} k_1 e^{-1t} \\ k_2 e^{-2t} \\ k_3 e^{3t} \end{bmatrix},$$

or

$$\begin{aligned} x_1(t) &= k_1 e^{-1t} + 2k_2 e^{-2t} + k_3 e^{3t} \\ x_2(t) &= \phantom{k_1 e^{-1t} + 2} -k_2 e^{-2t} + 2k_3 e^{3t} \\ x_3(t) &= -k_1 e^{-1t} + 2k_2 e^{-2t} + k_3 e^{3t} \end{aligned}.$$

Exercises

1. Two species live in the same habitat and compete for food. By observation $x_1(0) = 600$ and $x_2(0) = 200$, where $t = 0$ refers to the present. Future population is predicted to follow the equations

$$\begin{aligned} x_1'(t) &= -2x_1(t) + 4x_2(t) \\ x_2'(t) &= x_1(t) - 2x_2(t) \end{aligned},$$

where t is measured in years and $x_1(t)$ is the population of species 1 after t years and $x_2(t)$ is the population of species 2 after t years. Find $x_1(t)$ and $x_2(t)$ as functions of t. Will these species coexist indefinitely?

2. Solve the system

$$\begin{aligned} x_1' &= x_2 \\ x_2' &= x_3 \\ x_3' &= 8x_1 - 14x_2 + 7x_3 \end{aligned}.$$

3. Solve the system
$$\mathbf{x}'(t) = \begin{bmatrix} 2 & 1 & 1 \\ 1 & 2 & 1 \\ 1 & 1 & 2 \end{bmatrix} \mathbf{x}(t).$$

4. Solve the system
$$\mathbf{x}'(t) = \begin{bmatrix} 3 & 2 & 4 \\ 2 & 0 & 2 \\ 4 & 2 & 3 \end{bmatrix} \mathbf{x}(t).$$

Section 4.3 Powers of a Matrix and the Fibonacci Sequence

If we can diagonalize an $n \times n$ matrix A, we can find a formula for all the positive integral powers of A. For a diagonal matrix

$$D = \begin{bmatrix} d_1 & 0 & \cdots & 0 \\ 0 & d_2 & & \vdots \\ \vdots & & \ddots & 0 \\ 0 & \cdots & 0 & d_n \end{bmatrix},$$

we have

$$D^q = \begin{bmatrix} d_1^q & 0 & \cdots & 0 \\ 0 & d_2^q & & \vdots \\ \vdots & & \ddots & 0 \\ 0 & \cdots & 0 & d_n^q \end{bmatrix},$$

for any positive integer q. We also have

$$\begin{aligned}(P^{-1}AP)^q &= (P^{-1}AP)(P^{-1}AP)\cdots(P^{-1}AP) \\ &= P^{-1}A^q P,\end{aligned}$$

so that if
$$P^{-1}AP = D,$$

Applications

we have
$$P^{-1}A^qP = (P^{-1}AP)^q = D^q.$$
Thus
$$P^{-1}A^qP = D^q,$$
which is equivalent to
$$A^q = PD^qP^{-1}.$$

Procedure *To find A^q if A is diagonalizable: Find the eigenvalues $d_1, d_2, \cdots d_n$, (not necessarily distinct) of A and a matrix P such that $P^{-1}AP = D$, the diagonal matrix with $d_1, d_2, \cdots, d_n$ on the diagonal (in that order). Then*

$$A^q = P \begin{bmatrix} d_1^q & 0 & \cdots & 0 \\ 0 & d_2^q & & \vdots \\ \vdots & & \ddots & 0 \\ 0 & \cdots & 0 & d_n^q \end{bmatrix} P^{-1}.$$

In a book dating back to 1202, Leonardo Fibonacci defined the following sequence:
$$s_0 = 1, s_1 = 1,$$
and for $n \geq 1$,
$$s_{n+1} = s_n + s_{n-1}.$$
Thus
$$s_2 = s_1 + s_0 = 1 + 1 = 2,$$
$$s_3 = s_2 + s_1 = 2 + 1 = 3.$$

Each term is the sum of the previous two. The problem is to find a formula for s_n in terms of n, so that one may find s_{100} without finding all the preceding terms. Note that

$$\begin{bmatrix} 0 & 1 \\ 1 & 1 \end{bmatrix} \begin{bmatrix} s_{n-1} \\ s_n \end{bmatrix} = \begin{bmatrix} s_n \\ s_{n-1} + s_n \end{bmatrix} = \begin{bmatrix} s_n \\ s_{n+1} \end{bmatrix}.$$

Let
$$A = \begin{bmatrix} 0 & 1 \\ 1 & 1 \end{bmatrix}.$$

Then from the above we see that

$$A\begin{bmatrix} s_{n-1} \\ s_n \end{bmatrix} = \begin{bmatrix} s_n \\ s_{n+1} \end{bmatrix},$$

$$A^2\begin{bmatrix} s_{n-1} \\ s_n \end{bmatrix} = A\begin{bmatrix} s_n \\ s_{n+1} \end{bmatrix} = \begin{bmatrix} s_{n+1} \\ s_{n+2} \end{bmatrix},$$

and

$$A^q\begin{bmatrix} s_{n-1} \\ s_n \end{bmatrix} = \begin{bmatrix} s_{n+q-1} \\ s_{n+q} \end{bmatrix},$$

where q is a positive integer. Taking $n = 1$ in the above formula gives

$$A^q\begin{bmatrix} s_0 \\ s_1 \end{bmatrix} = \begin{bmatrix} s_q \\ s_{q+1} \end{bmatrix}.$$

If A is diagonalizable, we have a formula for the powers of A, so we proceed to see if we can find an invertible P such that $P^{-1}AP$ is diagonal. We have

$$|A - kI| = \begin{vmatrix} -k & 1 \\ 1 & 1-k \end{vmatrix} = k^2 - k - 1.$$

Using the quadratic formula, the eigenvalues of A are

$$k = \frac{1 \pm \sqrt{1+4}}{2} = \frac{1}{2} \pm \frac{\sqrt{5}}{2}.$$

Once we have found P such that

$$P^{-1}AP = \begin{bmatrix} \frac{1}{2} + \frac{\sqrt{5}}{2} & 0 \\ 0 & \frac{1}{2} - \frac{\sqrt{5}}{2} \end{bmatrix},$$

then

$$A^q = P\begin{bmatrix} \left(\frac{1}{2} + \frac{\sqrt{5}}{2}\right)^q & 0 \\ 0 & \left(\frac{1}{2} - \frac{\sqrt{5}}{2}\right)^q \end{bmatrix} P^{-1},$$

and

$$\begin{bmatrix} s_q \\ s_{q+1} \end{bmatrix} = A^q \begin{bmatrix} s_0 \\ s_1 \end{bmatrix} = A^q \begin{bmatrix} 1 \\ 1 \end{bmatrix}.$$

Applications

We leave the calculation of P to the reader. The resulting formula for the sequence is

$$s_q = \frac{1}{\sqrt{5}}\left(\left(\frac{1}{2}+\frac{\sqrt{5}}{2}\right)^{q+1} - \left(\frac{1}{2}-\frac{\sqrt{5}}{2}\right)^{q+1}\right).$$

If q is large, you may observe that $\left(\frac{1}{2}-\frac{\sqrt{5}}{2}\right)^{q+1}$ is relatively small, so we have the approximation

$$s_q \cong \frac{1}{\sqrt{5}}\left(\frac{1}{2}+\frac{\sqrt{5}}{2}\right)^{q+1}.$$

Exercises

1. Use a calculator or computer to find approximations for s_{100} and s_{1000} in the Fibonacci sequence, using the formula developed above.

2. Develop a formula for A^q if $A = \begin{bmatrix} 0 & 1 \\ 2 & 1 \end{bmatrix}$.

3. Use the matrix in problem 2 to develop a formula for the qth term of the sequence defined by $u_0 = u_1 = 1$, and $u_{n+1} = u_n + 2u_{n-1}$ for positive integers $n \geq 1$.

Section 4.4 Least Squares Approximation

We have seen that the matrix equation $A\mathbf{x} = \mathbf{b}$ has no solution when $\mathbf{b}$ is not in the column space of A. However we may wish to find an approximate solution, even if no exact solution exists. The equation $A^T A\mathbf{x} = A^T \mathbf{b}$, which is called the normal equation, always has a solution (unique if the columns of A are linearly independent). A solution $\mathbf{v}$ to the **normal equation** has the property that $A\mathbf{v}$ is the closest vector to $\mathbf{b}$ in the column space of A (closest in the sense that

$(A\mathbf{v} - \mathbf{b}) \cdot (A\mathbf{v} - \mathbf{b}) \leq (\mathbf{w} - \mathbf{b}) \cdot (\mathbf{w} - \mathbf{b})$ for all $\mathbf{w}$ in columnspace(A)). A solution $\mathbf{v}$ to the normal equation $A^T A \mathbf{x} = A^T \mathbf{b}$ is called a **least squares approximate solution** to $A\mathbf{x} = \mathbf{b}$, and $A\mathbf{v}$ is the projection of $\mathbf{b}$ on the column space of A.

Suppose we are given a collection of points $(x_1, y_1), (x_2, y_2), \cdots, (x_n, y_n)$, and we wish to find a line such that the sum of the squares of the distances of these points from the line is minimized. This is the line that best fits the data points. We will assume that no two of the points have the same x value, in other words, no two of the points lie on a vertical line. Let

$$A = \begin{bmatrix} 1 & x_1 \\ 1 & x_2 \\ \vdots & \vdots \\ 1 & x_n \end{bmatrix},$$

and

$$\mathbf{b} = \begin{bmatrix} y_1 \\ y_2 \\ \vdots \\ y_n \end{bmatrix}.$$

A line whose equation is $y = a_0 + a_1 x$ will go through all these points if and only if

$$\begin{bmatrix} 1 & x_1 \\ 1 & x_2 \\ \vdots & \vdots \\ 1 & x_n \end{bmatrix} \begin{bmatrix} a_0 \\ a_1 \end{bmatrix} = \begin{bmatrix} y_1 \\ y_2 \\ \vdots \\ y_n \end{bmatrix},$$

or equivalently

$$A \begin{bmatrix} a_0 \\ a_1 \end{bmatrix} = \mathbf{b}.$$

In other words to find a line going through all the points we must solve the above equation for a_0 and a_1. When there are more than two points they may not lie on a line, so an exact solution may not exist. Thus we solve the normal equation

$$A^T A \begin{bmatrix} a_0 \\ a_1 \end{bmatrix} = A^T \mathbf{b}.$$

Applications

Example *The observations of an experiment result in the following points: $(-1, -2)$, $(0, 1)$, $(1, 3)$, and $(2, 6)$. Find the line best fitting this data in the least squares sense. Solution:*

$$A = \begin{bmatrix} 1 & -1 \\ 1 & 0 \\ 1 & 1 \\ 1 & 2 \end{bmatrix},$$

and

$$\mathbf{b} = \begin{bmatrix} -2 \\ 1 \\ 3 \\ 6 \end{bmatrix}.$$

$$A^T A = \begin{bmatrix} 1 & 1 & 1 & 1 \\ -1 & 0 & 1 & 2 \end{bmatrix} \begin{bmatrix} 1 & -1 \\ 1 & 0 \\ 1 & 1 \\ 1 & 2 \end{bmatrix}$$

$$= \begin{bmatrix} 4 & 2 \\ 2 & 6 \end{bmatrix},$$

$$A^T \mathbf{b} = \begin{bmatrix} 1 & 1 & 1 & 1 \\ -1 & 0 & 1 & 2 \end{bmatrix} \begin{bmatrix} -2 \\ 1 \\ 3 \\ 6 \end{bmatrix}$$

$$= \begin{bmatrix} 8 \\ 17 \end{bmatrix}.$$

Thus the normal equation is

$$\begin{bmatrix} 4 & 2 \\ 2 & 6 \end{bmatrix} \begin{bmatrix} a_0 \\ a_1 \end{bmatrix} = \begin{bmatrix} 8 \\ 17 \end{bmatrix},$$

and has solutions $a_0 = .7$, and $a_1 = 2.6$, so that

$$y = .7 + 2.6x$$

is the equation of the desired line.

Exercises

1. Find the least sqare fit line for the observations $(0, 5)$, $(1, 4)$, $(2, 8)$, $(3, 13)$, $(4, 15)$.

2. Let
$$A = \begin{bmatrix} 1 & -1 & 1 & -1 \\ 1 & 0 & 0 & 0 \\ 1 & 1 & 1 & 1 \\ 1 & 2 & 4 & 8 \\ 1 & 3 & 9 & 27 \end{bmatrix}, \text{ and } \mathbf{b} = \begin{bmatrix} -13 \\ 2 \\ 1 \\ -4 \\ 20 \end{bmatrix}.$$
$A\mathbf{x} = \mathbf{b}$ does not have a solution. Using a computer solve the normal equation for the least squares approximate solution.

Chapter 5

Geometry

Section 5.1 Cross Product and Planes in R^3

In R^3 a **plane through the origin** is the solution set of an equation of the form $ax + by + cz = 0$, where at least one of a, b, c is not zero. In other words a plane through the origin is the null space of the matrix $\begin{bmatrix} a & b & c \end{bmatrix}$, or the set of all vectors orthogonal to (a, b, c). Since a, b, and c are not all zero, this subspace has dimension 2. We can find a basis ot this space by our usual method of finding a basis of the null space.

Suppose we want to go the other way. Suppose $\mathbf{u} = (a_1, a_2, a_3)$ and $\mathbf{v} = (b_1, b_2, b_3)$ are linearly independent vectors and we want to find the equation of the plane they span. We need to find a vector orthogonal to both $\mathbf{u}$ and $\mathbf{v}$. We can do this by finding a basis of the null space of the matrix

$$\begin{bmatrix} a_1 & a_2 & a_3 \\ b_1 & b_2 & b_3 \end{bmatrix}.$$

Example   Find the equation of the plane spanned by the vectors $(1, 3, 1)$

and $(2,1,0)$. Solution: Form the matrix
$$\begin{bmatrix} 1 & 3 & 1 \\ 2 & 1 & 0 \end{bmatrix}.$$

The RREF is
$$\begin{bmatrix} 1 & 0 & -\frac{1}{5} \\ 0 & 1 & \frac{2}{5} \end{bmatrix},$$

and basis of the null space is $(-1, 2, -5)$. Note that this vector is orthogonal to both of the given vectors. Thus the equation of the plane is $-x + 2y - 5z = 0$ or equivalently $x - 2y + 5z = 0$.

In R^3 there is another way to find a vector orthogonal to two given vectors **u** and **v**. It is the cross product of **u** and **v**. The **cross product** of **u** and **v** is a vector orthogonal to both **u** and **v** and given by the formula

$$\mathbf{u} \times \mathbf{v} = (a_2 b_3 - a_3 b_2,\ a_3 b_1 - a_1 b_3,\ a_1 b_2 - a_2 b_1),$$

where $\mathbf{u} = (a_1, a_2, a_3)$, and $\mathbf{v} = (b_1, b_2, b_3)$.

An easy way to remember this is $\mathbf{u} \times \mathbf{v} = \begin{vmatrix} \mathbf{i} & \mathbf{j} & \mathbf{k} \\ a_1 & a_2 & a_3 \\ b_1 & b_2 & b_3 \end{vmatrix}$, which when we expand by minors around the first row gives

$$\mathbf{u} \times \mathbf{v} = \begin{vmatrix} a_2 & a_3 \\ b_2 & b_3 \end{vmatrix} \mathbf{i} - \begin{vmatrix} a_1 & a_3 \\ b_1 & b_3 \end{vmatrix} \mathbf{j} + \begin{vmatrix} a_1 & a_2 \\ b_1 & b_2 \end{vmatrix} \mathbf{k}.$$

For this to give $\mathbf{u} \times \mathbf{v}$, we interpret the letters **i**, **j**, and **k** as the natural basis vectors of R^3.

Example Find a vector orthogonal to both $(1, 3, 1)$ and $(2, 1, 0)$. Solution:

$$\begin{vmatrix} \mathbf{i} & \mathbf{j} & \mathbf{k} \\ 1 & 3 & 1 \\ 2 & 1 & 0 \end{vmatrix} = \mathbf{i}(0 - 1) - \mathbf{j}(0 - 2) + \mathbf{k}(1 - 6)$$
$$= -1\mathbf{i} + 2\mathbf{j} - 5\mathbf{k} = (-1, 2, -5).$$

Compare with the previous example.

Remark $\begin{vmatrix} \mathbf{i} & \mathbf{j} & \mathbf{k} \\ 1 & 3 & 1 \\ 2 & 1 & 0 \end{vmatrix}$ *is not really a determinant, since* $\mathbf{i}$, $\mathbf{j}$, *and* $\mathbf{k}$ *are vectors, and a determinant does not have vectors as individual entries. To avoid this logical difficulty we could just leave the first row blank and compute the cofactors of the first row as the components of* $\mathbf{u} \times \mathbf{v}$, *but the use of* $\mathbf{i}$, $\mathbf{j}$, *and* $\mathbf{k}$ *is convenient.*

For fixed a, b, c let π denote the plane $ax + by + cz = 0$. Then the equation $ax + by + cz = d \neq 0$ is a plane parallel to π but not through the origin. As d varies this gives rise to the family of all planes parallel to π.

Example Find the equation of the plane orthogonal to the vector $(3, -1, 4)$ and through the point $(2, 1, 0)$. Solution: The plane parallel to this and through the origin is $3x - y + 4z = 0$, so the equation of this plane will be $3x - y + 4z = d$ for some d. To find d, we substitute in the values of x, y, and z for the point we know is on the plane, namely $(2, 1, 0)$. Thus we get $3 \cdot 2 - 1 + 4 \cdot 0 = d$, or $d = 5$. So the equation of the plane is $3x - y + 4z = 5$.

Example Find the equation of the plane through the points $(1, 1, 2)$, $(1, 3, 0)$, and $(2, -1, 2)$. Solution: If these triples are solutions of $ax + by + cz = d$, the difference of any two of them is a solution of $ax + by + cz = 0$, for the same a, b, and c. So we form $\mathbf{u} = (1, 3, 0) - (1, 1, 2) = (0, 2, -2)$, and $\mathbf{v} = (2, -1, 2) - (1, 1, 2) = (1, -2, 0)$. The cross product $\mathbf{u} \times \mathbf{v}$ will now give a triple that we can use for a, b, and c. $(0, 2, -2) \times (1, -2, 0) = (-4, -2, -2)$, so we know that the equation we are seeking is $-4x - 2y - 2z = d$, for some value of d. To find d, we substitute into the equation the components of any of the given points, for example using the point $(1, 3, 0)$ we get $-4 \cdot 1 - 2 \cdot 3 - 2 \cdot 0 = d$, so $d = -10$, and the equation we are seeking is $-4x - 2y - 2z = -10$, or equivalently $4x + 2y + 2z = 10$.

There are other ways of finding the equation of the plane through 3 given points, but the one just demonstrated using the cross product is probably the quickest and easiest. It is interesting to note however that the equation of the plane through the 3 points $p = (p_1, p_2, p_3)$, $q = (q_1, q_2, q_3)$, and $r = (r_1, r_2, r_3)$ not on a line is given by

$$\begin{vmatrix} x & y & z & 1 \\ p_1 & p_2 & p_3 & 1 \\ q_1 & q_2 & q_3 & 1 \\ r_1 & r_2 & r_3 & 1 \end{vmatrix} = 0.$$

Remark *The three points p, q, and r lie on a line if $(q-p)$ is multiple of $(r-p)$, and in this case the above determinant is zero for all values of x, y, and z, and so does not give the equation of a plane.*

The length of $\mathbf{u} \times \mathbf{v}$ can be shown to satisfy

$$|\mathbf{u} \times \mathbf{v}| = |\mathbf{u}||\mathbf{v}||\sin \Phi|,$$

where Φ is the angle between $\mathbf{u}$ and $\mathbf{v}$.

Fact $\mathbf{u} \times \mathbf{v}$ *gives a vector whose length is the area of the parallelogram with edges $\mathbf{u}$ and $\mathbf{v}$. One half of $|\mathbf{u} \times \mathbf{v}|$ is the area of the triangle with two sides $\mathbf{u}$ and $\mathbf{v}$.*

If we have 3 vectors $\mathbf{w}$, $\mathbf{u}$, and $\mathbf{v}$, we can form $\mathbf{w} \cdot (\mathbf{u} \times \mathbf{v})$. This is called the **triple scalar product** and we have

Fact

$$\mathbf{w} \cdot (\mathbf{u} \times \mathbf{v}) = \begin{vmatrix} \mathbf{w} \\ \mathbf{u} \\ \mathbf{v} \end{vmatrix},$$

the determinant of the matrix with rows $\mathbf{w}$, $\mathbf{u}$, and $\mathbf{v}$.

It can be shown that $\mathbf{w} \cdot \mathbf{z} = |\mathbf{w}||\mathbf{z}|(\cos \tau)$, where τ is the angle between $\mathbf{w}$ and $\mathbf{z}$. Using this and facts above about the cross product, one can show that

Fact

$\mathbf{w} \cdot (\mathbf{u} \times \mathbf{v}) = \pm$ *volume of a parallelapiped with edges $\mathbf{w}$, $\mathbf{u}$, and $\mathbf{v}$.*

Where the + occurs if the angle between $\mathbf{w}$ and $\mathbf{u} \times \mathbf{v}$ is acute ($\mathbf{w}$ and $\mathbf{u} \times \mathbf{v}$ are both on the same side of the plane determined by $\mathbf{u}$ and $\mathbf{v}$, in this case we say $\mathbf{w}, \mathbf{u}, \mathbf{v}$ is a right handed system) and the $-$ occurs if the angle between $\mathbf{w}$ and $\mathbf{u} \times \mathbf{v}$ is obtuse.

Geometry

Exercises

1. Find the equation of the plane through the origin and orthogonal to the vector $(1, 5, -1)$.

2. Find the equation of the plane parallel to the plane in question 1 but going through the point $(2, 2, 0)$.

3. Find the equation of the plane through the points $(1, 1, 3)$, $(1, 4, 1)$, and $(1, 2, 2)$.

4. Find a vector orthogonal to both $(1, -2, 4)$ and $(1, 0, 1)$.

Section 5.2 Lines in R^3

A **line** in R^3 can be described as the intersection of two planes.

Example *Find the intersection of the planes $3x+2y-z = 3$, and $2x+3y+2z = 5$. Solution: To find the intersection, we solve the equations simultaneously. The augmented matrix of this system of equations is*

$$\begin{bmatrix} 3 & 2 & -1 & 3 \\ 2 & 3 & 2 & 5 \end{bmatrix},$$

with RREF

$$\begin{bmatrix} 1 & 0 & -\frac{7}{5} & -\frac{1}{5} \\ 0 & 1 & \frac{8}{5} & \frac{9}{5} \end{bmatrix}.$$

The complete set of solutions is given by

$$\begin{aligned} x &= -\frac{1}{5} + \frac{7}{5}t \\ y &= \frac{9}{5} - \frac{8}{5}t \\ z &= t. \end{aligned}$$

This shows another way that a line can be described, namely by a system of **parametric equations** of the form

$$\begin{aligned} x &= at + x_0 \\ y &= bt + y_0, \\ z &= ct + z_0 \end{aligned}$$

where (a, b, c) is a vector in the direction of the line and (x_0, y_0, z_0) is a point on the line.

When (a, b, c) is a vector in the direction of a line L, the numbers a, b, c are called **direction numbers of** L. They are not unique, since if (a, b, c) is in the direction of L, so is (qa, qb, qc) for any number q.

Remark *The system of parametric equations describing a line is not unique. Any set of direction numbers can be used as the coefficients of the parameter, and the coordinates of any point on the line can be used in place of x_0, y_0, and z_0.*

Example *Find a system of parametric equations describing the line in the direction of the vector $(3, -2, 4)$ and through the point $(5, 1, -8)$.*
Solution:
$$\begin{aligned} x &= 3t + 5 \\ y &= -2t + 1 \\ z &= 4t - 8 \end{aligned}.$$

Example *Find a system of parametric equations describing the line determined by the points $(4, -6, 3)$ and $(2, 8, 5)$.* Solution: A vector in the direction of the line is given by $(2, 8, 5) - (4, -6, 3) = (-2, 14, 2)$, and we can use either point as the point on the line, thus the line is described by
$$\begin{aligned} x &= -2t + 2 \\ y &= 14t + 8 \\ z &= 2t + 5 \end{aligned}.$$

If a line is described by a system of parametric equations and we wish to find two planes that intersect in the line, we can do so as in the following example.

Example *Find two planes that intersect in the line described by*
$$\begin{aligned} x &= 2t - 3 \\ y &= 7t + 4 \\ z &= t - 5 \end{aligned}.$$

Solution: Solving each of the equations for the parameter t we get $t = \frac{x+3}{2}$, $t = \frac{y-4}{7}$, $t = z+5$. Thus a point on this line must satisfy

$$\frac{x+3}{2} = \frac{y-4}{7},$$

and

$$\frac{y-4}{7} = z+5.$$

Simplifying these equations we have

$$\begin{aligned} 7x - 2y &= -29 \\ y - 7z &= 39 \end{aligned}.$$

These are the equations of two planes that intersect in the given line as desired. (Note: the first plane is parallel to the z axis and the second is parallel to the X axis. We could also have used the plane $\frac{x+3}{2} = z+5$, or $x - 2z = 7$ that is parallel to the Y axis. The intersection of any two of these three planes is the given line.)

Example Find two planes that intersect in the line described by

$$\begin{aligned} x &= 3 \\ y &= 1 + 2t \\ z &= -4 + 3t \end{aligned}.$$

Solution: The equation $x = 3$ describes a plane parallel to the YZ plane. Then solving the remaining two equations for t and equating, we get $\frac{y-1}{2} = \frac{z+4}{3}$, or $3y - 2z = 11$. The intersection of these two planes is the given line.

In R^3 two lines are called **skew** if they are not parallel and also do not intersect.

Warning When finding the intersection of two lines described by parametric equations it is important to use a different letter as the parameter in the two sets of equations, since the point of intersection may not occur at the same value of the parameter for both lines.

Example Find the intersection of the lines

$$x = 3t - 4 \qquad x = -t + 3$$
$$y = -2t + 6 \qquad y = 4t - 2$$
$$z = t + 1 \qquad z = 2t + 1$$

Solution: Change the name of the parameter in the first system to s, so that we have

$$x = 3s - 4 \qquad x = -t + 3$$
$$y = -2s + 6 \qquad y = 4t - 2$$
$$z = s + 1 \qquad z = 2t + 1$$

Now a point of intersection corresponds to a pair of values of s and t that give the same point. Thus we get the system of 3 equations and 2 unknowns

$$3s - 4 = -t + 3$$
$$-2s + 6 = 4t - 2 \ .$$
$$s + 1 = 2t + 1$$

This system has solution $s = 2$, $t = 1$, so the point $(2, 2, 3)$ is on both lines.

Example Find the intersection of the lines described by

$$x = 2t + 5 \qquad x = 3t - 2$$
$$y = 4t + 1 \qquad y = 6t + 3 \ .$$
$$z = 8t - 5 \qquad z = 12t - 7$$

Solution: These lines are parallel. Looking at the coefficients of the parameters we see that $(3, 6, 12) = \frac{3}{2}(2, 4, 8)$, so the lines are in the same direction. Parallel lines do not intersect, so there is no point of intersection.

Example Find the point of intersection of the lines described by

$$x = 3s - 4 \qquad x = -t + 3$$
$$y = -2s + 6 \qquad y = 4t - 2 \ .$$
$$z = s + 1 \qquad z = 2t - 1$$

Geometry

Solution: We look for a pair of values s and t that will give the same point. We get the following system of 3 equations in two unknowns

$$3s - 4 = -t + 3$$
$$-2s + 6 = 4t - 2$$
$$s + 1 = 2t - 1$$

or

$$3s + t = 7$$
$$-2s - 4t = -8 \ .$$
$$s - 2t = -2$$

The augmented matrix of this system is

$$\begin{bmatrix} 3 & 1 & 7 \\ -2 & -4 & -8 \\ 1 & -2 & -2 \end{bmatrix},$$

which has RREF

$$\begin{bmatrix} 1 & 0 & 0 \\ 0 & 1 & 0 \\ 0 & 0 & 1 \end{bmatrix},$$

showing that there is no solution. Thus the two lines do not intersect.

Exercises

1. Find the direction numbers of the folowing lines:

$$L_1: \begin{array}{l} x = 2t + 3 \\ y = t + 1 \\ z = 5 \end{array} \qquad L_2: \begin{array}{l} x = t + 1 \\ y = 2t + 2 \\ z = 5t + 3 \end{array}$$

$$L_3: \begin{array}{l} x = -t + 1 \\ y = -t + 1 \\ z = -t + 1 \end{array}.$$

2. Find parametric equations describing a line through the points $(1, 1, 1)$ and $(2, 3, 4)$.

3. Find parametric equations describing a line though the point $(8, 7, 6)$ and parallel to the line

$$L: \begin{aligned} x &= t + 1 \\ y &= 3t + 2 \\ z &= 5t + 3 \end{aligned}.$$

4. Find parametric equations of the line through the points $(5, 3, 4)$ and $(6, 1, 1)$.

5. For each pair of lines determine if they are parallel, skew, or intersecting. If they intersect, find the point of intersection.

   (a)
   $$\begin{aligned} x &= -3t + 1 \\ y &= 2t + 2 \\ z &= -t + 3 \end{aligned} \quad \text{and} \quad \begin{aligned} x &= t + 4 \\ y &= 2t + 5 \\ z &= 3t + 6 \end{aligned}$$

   (b)
   $$\begin{aligned} x &= -3t + 1 \\ y &= 2t + 2 \\ z &= -t + 3 \end{aligned} \quad \text{and} \quad \begin{aligned} x &= 6t + 3 \\ y &= -4t + 3 \\ z &= 2t + 4 \end{aligned}$$

   (c)
   $$\begin{aligned} x &= t + 4 \\ y &= 2t + 5 \\ z &= 3t + 6 \end{aligned} \quad \text{and} \quad \begin{aligned} x &= -3t + 8 \\ y &= 2t + 5 \\ z &= 6t + 3 \end{aligned}$$

   (d)
   $$\begin{aligned} x &= -3t + 8 \\ y &= 2t + 5 \\ z &= 6t + 3 \end{aligned} \quad \text{and} \quad \begin{aligned} x &= -2t + 7 \\ y &= -t + 8 \\ z &= -t + 10 \end{aligned}$$

Geometry

Section 5.3 Projection

The **projection of a vector u on a vector v** is given by
$$proj_{\mathbf{v}}\mathbf{u} = \left(\frac{\mathbf{u} \cdot \mathbf{v}}{\mathbf{v} \cdot \mathbf{v}}\right)\mathbf{v}.$$
This is a vector in the direction of **v** with length given by
$$|proj_{\mathbf{v}}\mathbf{u}| = |\mathbf{u}||\cos\Phi|,$$
where Φ is the angle between **u** and **v**.

Fact $(\mathbf{u} - proj_{\mathbf{v}}\mathbf{u})$ *is orthogonal to* **v**.

Reason.
$$\begin{aligned}(\mathbf{u} - proj_{\mathbf{v}}\mathbf{u}) \cdot \mathbf{v} &= \left(\mathbf{u} - \left(\frac{\mathbf{u} \cdot \mathbf{v}}{\mathbf{v} \cdot \mathbf{v}}\right)\mathbf{v}\right) \cdot \mathbf{v} \\ &= \mathbf{u} \cdot \mathbf{v} - \left(\frac{\mathbf{u} \cdot \mathbf{v}}{\mathbf{v} \cdot \mathbf{v}}\right)(\mathbf{v} \cdot \mathbf{v}) = 0.\end{aligned}$$

Thus we can write
$$\mathbf{u} = proj_{\mathbf{v}}\mathbf{u} + (\mathbf{u} - proj_{\mathbf{v}}\mathbf{u}) = c\mathbf{v} + \mathbf{w},$$
where **w** is orthogonal to **v**.

This decomposition is unique, since
$$\mathbf{u} \cdot \mathbf{v} = \mathbf{u} \cdot (c\mathbf{v} + \mathbf{w}) = c(\mathbf{v} \cdot \mathbf{v})$$
forcing
$$c = \frac{\mathbf{u} \cdot \mathbf{v}}{\mathbf{v} \cdot \mathbf{v}}.$$

The **projection of a vector u on a line** L can be taken as
$$proj_L\mathbf{u} = proj_{\mathbf{v}}\mathbf{u},$$
where **v** is any vector in the direction of L.

Given a plane π through the origin in R^3 and a vector **u** not in π, we can resolve **u** as a sum $\mathbf{v} + \mathbf{w}$ where **v** is in π and **w** is orthogonal (perpendicular) to π. This decomposition is unique and **v** is called the **projection of u on a plane** π. In this decomposition **w** is $proj_{\mathbf{n}}\mathbf{u}$, where **n** is any vector orthogonal to π. Thus we have
$$proj_{\pi}\mathbf{u} = \mathbf{u} - proj_{\mathbf{n}}\mathbf{u}.$$

Example Find the projection of $\mathbf{u} = (5, 3, 6)$ on the line
$$L: \begin{array}{rl} x = & t + 3 \\ y = & 2t - 1 \\ z = & -t + 4 \end{array}.$$

Solution: $\mathbf{v} = (1, 2, -1)$ is a vector in the direction of L. Therefore
$$proj_L \mathbf{u} = proj_{\mathbf{v}} \mathbf{u}$$
$$= \left(\frac{(5, 3, 6) \cdot (1, 2, -1)}{(1, 2, -1) \cdot (1, 2, -1)} \right) (1, 2, -1)$$
$$= \frac{5}{6}(1, 2, -1).$$

Example Find the projection of $\mathbf{u} = (5, 3, 6)$ on the plane π given by $x + 2y - z = 0$. Solution: The vector $\mathbf{n} = (1, 2, -1)$ is orthogonal to π, therefore
$$proj_\pi \mathbf{u} = \mathbf{u} - proj_\mathbf{n} \mathbf{u}$$
$$= (5, 3, 6) - \left(\frac{(5, 3, 6) \cdot (1, 2, -1)}{(1, 2, -1) \cdot (1, 2, -1)} \right) (1, 2, -1)$$
$$= (5, 3, 6) - \frac{5}{6}(1, 2, -1) = (25/6, 8/6, 41/6).$$

Fact If π is a plane given by the equation $ax + by + cz = d$ and P is a point with coordinates x_1, y_1, z_1, then the distance D from P to π is given by
$$D = \frac{|ax_1 + by_1 + cz_1 - d|}{\sqrt{a^2 + b^2 + c^2}}$$

Remark If the plane goes through the origin ($d = 0$), you can verify that D is just the length of the projection of (x_1, y_1, z_1) on (a, b, c).

Exercises

1. Find the projection of the vector $(1, 2, 3)$ on the line described by
$$\begin{array}{rl} x = & t + 1 \\ y = & t - 1 \\ z = & 2t + 2 \end{array}.$$

2. Find the projection of the vector $(1, 3, -2)$ on the plane $2x - y + z = 0$.

3. Find the distance from the point $(2, 3, 4)$ to the XY plane.

4. Find the distance from $(2, 3, 4)$ to the plane $x + 2y + 3z = 6$.

5. Show that there exist two planes whose distance from the point $(2, 3, 4)$ is the same as the distance determined in problem 2.

6.
   (a) Find the distance from the point $(5, 3, 6)$ to the plane $\pi_1 : x + 2y - z = 0$, and find the distance of the same point to the plane $\pi_2 : x + 2y - z = 10$.

   (b) Find the distance between π_1 and π_2.

   (c) Is the point $(5, 3, 6)$ between the planes π_1 and π_2?

Section 5.4 Orthogonal Matrices and Quadrics

We are interested in linear transformations that preserve length of vectors and angles between vectors.

Fact *A linear transformation L from R^n to R^n preserves both length and angle if and only if it preserves the dot product.*

Reason. The length of $\mathbf{u}$ is given by

$$|\mathbf{u}| = \sqrt{\mathbf{u} \cdot \mathbf{u}},$$

and

$$\cos \Phi = \frac{u \cdot v}{|u|\,|v|},$$

so clearly if $L(\mathbf{u}) \cdot L(\mathbf{v}) = \mathbf{u} \cdot \mathbf{v}$ for all choices of $\mathbf{u}$ and $\mathbf{v}$, then L preserves lengths and angles.

Conversely, just preserving length is enough to force L to preserve the dot product, and hence also angles. To see this note that if $|L(\mathbf{u})| = |\mathbf{u}|$ for all vector $\mathbf{u}$, then $L(\mathbf{u}) \cdot L(\mathbf{u}) = \mathbf{u} \cdot \mathbf{u}$ for all $\mathbf{u}$, and hence

$$L(\mathbf{u}+\mathbf{v}) \cdot L(\mathbf{u}+\mathbf{v}) = (\mathbf{u}+\mathbf{v}) \cdot (\mathbf{u}+\mathbf{v})$$

for all $\mathbf{u}$ and $\mathbf{v}$. From this it follows that

$$L(\mathbf{u}) \cdot L(\mathbf{u}) + 2L(\mathbf{u}) \cdot L(\mathbf{v}) + L(\mathbf{v}) \cdot L(\mathbf{v}) = \mathbf{u} \cdot \mathbf{u} + 2\mathbf{u} \cdot \mathbf{v} + \mathbf{v} \cdot \mathbf{v},$$

and this implies $L(\mathbf{u}) \cdot L(\mathbf{v}) = \mathbf{u} \cdot \mathbf{v}$.

As we know, every linear transformation L from R^n to R^n is of the form $L(\mathbf{u}) = P\mathbf{u}$ for some $n \times n$ matrix P.

Fact *The linear transformation $L(\mathbf{u}) = P\mathbf{u}$ preserves dot products if and only if the matrix P is orthogonal.*

Reason. If we assume a the vectors $\mathbf{u}$, $\mathbf{v}$, $L(\mathbf{u})$, and $L(\mathbf{v})$ are written as rows, then $\mathbf{u}^T$, $\mathbf{v}^T$, $L(\mathbf{u})^T$, and $L(\mathbf{v})^T$ are column vectors, and $\mathbf{u} \cdot \mathbf{v} = \mathbf{u}^T \mathbf{v}$. If P is an orthogonal matrix $P^T P = I$. Thus we have

$$\begin{aligned} L(\mathbf{u}) \cdot L(\mathbf{v}) &= P\mathbf{u} \cdot P\mathbf{v} \\ &= (P\mathbf{u})^T P\mathbf{v} \\ &= \mathbf{u}^T P^T P \mathbf{v} \\ &= \mathbf{u}^T I \mathbf{v} = \mathbf{u}^T \mathbf{v} = \mathbf{u} \cdot \mathbf{v}. \end{aligned}$$

Conversely, the columns of P are the images of the standard basis vectors. Thus if L preserves dot products, the columns of P are orthonormal, and so P is an orthogonal matrix.

Fact *The only possible real eigenvalues for an orthogonal matrix P are ± 1.*

Reason. Since P preserves length, if $P\mathbf{v} = k\mathbf{v}$, we must have $|k\mathbf{v}| = |\mathbf{v}|$, and thus $k = \pm 1$.
$$L(\mathbf{u}) \cdot L(\mathbf{v}) = (L(\mathbf{u}))^T L(\mathbf{v}) = (P\mathbf{u})^T P\mathbf{v} = \mathbf{u}^T P^T P \mathbf{v} = \mathbf{u}^T \mathbf{v} = \mathbf{u} \cdot \mathbf{v}.$$

Fact *The orthogonal 2×2 matrices are exactly those of the form*

Geometry

$$\begin{bmatrix} \cos\Phi & \sin\Phi \\ -\sin\Phi & \cos\Phi \end{bmatrix} \quad \text{or} \quad \begin{bmatrix} \cos\Phi & \sin\Phi \\ \sin\Phi & -\cos\Phi \end{bmatrix}.$$

Reason. If the matrix

$$P = \begin{bmatrix} p_{11} & p_{12} \\ p_{21} & p_{22} \end{bmatrix}$$

is orthogonal, then

$$p_{11}^2 + p_{12}^2 = 1 = p_{21}^2 + p_{22}^2.$$

Let $p_{11} = \cos\Phi$. Since $-1 \le p_{11} \le 1$, this is always possible. Then

$$p_{12}^2 = 1 - p_{11}^2 = 1 - \cos^2\Phi = \sin^2\Phi,$$

Thus

$$p_{12} = \pm\sin\Phi.$$

If $p_{12} = -\sin\Phi$, then change Φ to $-\Phi$, and we have $\cos(-\Phi) = \cos\Phi = p_{11}$ while $\sin(-\Phi) = -\sin(\Phi) = p_{12}$. Thus we can assume there is an angle τ such that $p_{11} = \cos\tau$ and $p_{12} = \sin\tau$, and similarly we can assume there is an angle ω such that $p_{22} = \cos\omega$ and $p_{21} = \sin\omega$. Now using the fact that the rows of P are orthogonal vectors, we have

$$\begin{aligned} 0 &= p_{11}p_{21} + p_{12}p_{22} \\ &= (\cos\tau)(\sin\omega) + (\sin\tau)(\cos\omega) = \sin(\tau+\omega). \end{aligned}$$

Hence $\tau + \omega = 0$ or $\tau + \omega = \pi$. If $\tau + \omega = 0$, we get $\omega = -\tau$, and

$$P = \begin{bmatrix} \cos\tau & \sin\tau \\ -\sin\tau & \cos\tau \end{bmatrix}.$$

If $\tau + \omega = \pi$, then $\omega = \pi - \tau$, and

$$p_{21} = \sin\omega = \sin(\pi - \tau) = \sin\tau,$$

while

$$p_{22} = \cos\omega = \cos(\pi - \tau) = -\cos\tau,$$

and we have

$$P = \begin{bmatrix} \cos\tau & \sin\tau \\ \sin\tau & -\cos\tau \end{bmatrix}.$$

Remark *A linear transformation L always leaves the zero vector fixed, $L(\mathbf{O}) = \mathbf{O}$, so translations of the plane are not linear transformations.*

From the fact above we see that in R^2 there are two kinds of linear transformations that preserve the dot product (and therefore lengths and angles). We now show that the first type corresponds to a rotation of the plane (or the identity matrix), and the second corresponds to reflection about a line. They are easily distinguished because the matrix of the first type has determinant 1, and the second type has determinant -1.

Rotation
Let
$$L(\mathbf{u}) = P\mathbf{u} = \begin{bmatrix} \cos\Phi & \sin\Phi \\ -\sin\Phi & \cos\Phi \end{bmatrix} \mathbf{u}.$$
Then
$$\begin{vmatrix} \cos\Phi & \sin\Phi \\ -\sin\Phi & \cos\Phi \end{vmatrix} = \cos^2\Phi + \sin^2\Phi = 1,$$
and
$$\begin{vmatrix} \cos\Phi - k & \sin\Phi \\ -\sin\Phi & \cos\Phi - k \end{vmatrix} = k^2 - 2(\cos\Phi)k + 1.$$
Solving for k gives
$$k = \frac{2\cos\Phi \pm \sqrt{4\cos^2\Phi - 4}}{2},$$
which is not a real number unless $\cos^2\Phi = 1$, or $\cos\Phi = \pm 1$. If $\cos\Phi = 1$, $P = I$. If $\cos\Phi = -1$, $P = -I$ and L sends every vector to its negative, which is a rotation by 180 degrees. For $\cos^2\Phi \neq 1$, there are no real eigenvalues, hence no vector other than zero is fixed by L and so L is a pure rotation.

Reflection
Let
$$L(\mathbf{u}) = P\mathbf{u} = \begin{bmatrix} \cos\Phi & \sin\Phi \\ \sin\Phi & -\cos\Phi \end{bmatrix} \mathbf{u}.$$

Then
$$\begin{vmatrix} \cos \Phi & \sin \Phi \\ \sin \Phi & -\cos \Phi \end{vmatrix} = -\cos^2 \Phi - \sin^2 \Phi = -1,$$
and
$$\begin{vmatrix} \cos \Phi - k & \sin \Phi \\ \sin \Phi & -\cos \Phi - k \end{vmatrix} = k^2 - 1,$$
which has roots ± 1, so the second type always has eigenvalues 1 and -1. Thus there are orthogonal eigenvectors $\mathbf{v}$ and $\mathbf{w}$ with the property that $L(\mathbf{v}) = \mathbf{v}$, and $L(\mathbf{w}) = -\mathbf{w}$. Thus L is a reflection through a line in the direction of the vector $\mathbf{v}$.

Quadrics
An equation of the form $ax^2 + by^2 + cz^2 + 2dxy + 2exz + 2fyz = g$, represents a configuration in R^3 known as a **quadric surface**. This can be written in matrix form as

$$\begin{bmatrix} x & y & z \end{bmatrix} \begin{bmatrix} a & d & e \\ d & b & f \\ e & f & c \end{bmatrix} \begin{bmatrix} x \\ y \\ z \end{bmatrix} = \begin{bmatrix} g \end{bmatrix}.$$

There are other 3×3 matrices we could have used, but we choose the only real symmetric matrix that will work since we know something about real symmetric matrices, namely that they can be orthogonally diagonalized. Let

$$M = \begin{bmatrix} a & d & e \\ d & b & f \\ e & f & c \end{bmatrix}.$$

There exists an orthogonal 3×3 matrix P such that

$$P^{-1} M P = \begin{bmatrix} s_1 & 0 & 0 \\ 0 & s_2 & 0 \\ 0 & 0 & s_3 \end{bmatrix} = Q,$$

where s_1, s_2, and s_3 are the eigenvalues of M. Let

$$L \begin{bmatrix} x \\ y \\ z \end{bmatrix} = P^T \begin{bmatrix} x \\ y \\ z \end{bmatrix} = \begin{bmatrix} x' \\ y' \\ z' \end{bmatrix},$$

or equivalently
$$\begin{bmatrix} x \\ y \\ z \end{bmatrix} = P \begin{bmatrix} x' \\ y' \\ z' \end{bmatrix}.$$

Transposing both sides
$$\begin{bmatrix} x & y & z \end{bmatrix} = \begin{bmatrix} x' & y' & z' \end{bmatrix} P^T.$$

Substituting these last two equations in the equation
$$\begin{bmatrix} x & y & z \end{bmatrix} \begin{bmatrix} a & d & e \\ d & b & f \\ e & f & c \end{bmatrix} \begin{bmatrix} x \\ y \\ z \end{bmatrix} = \begin{bmatrix} g \end{bmatrix}$$

gives
$$\begin{bmatrix} x' & y' & z' \end{bmatrix} P^T M P \begin{bmatrix} x' \\ y' \\ z' \end{bmatrix} = \begin{bmatrix} g \end{bmatrix},$$

which becomes
$$\begin{bmatrix} x' & y' & z' \end{bmatrix} \begin{bmatrix} s_1 & 0 & 0 \\ 0 & s_2 & 0 \\ 0 & 0 & s_3 \end{bmatrix} \begin{bmatrix} x' \\ y' \\ z' \end{bmatrix} = \begin{bmatrix} g \end{bmatrix},$$

or
$$s_1(x')^2 + s_2(y')^2 + s_3(z')^2 = g.$$

We can think of this as a change of coordinates, or as moving the original axes to "standard position" where graphing becomes manageable.

Orthogonal Transformations on R^3

What can be said about a 3×3 orthogonal matrix P? The characteristic equation is a cubic, so at least one eigenvalue must be real. In fact the only possibilities are one real eigenvalue or three real eigenvalues. If $P\mathbf{v} = k\mathbf{v}$, then k must be ± 1 to preserve length. In either case if π is the plane perpendicular to $\mathbf{v}$, then the linear transformation $L(\mathbf{u}) = P\mathbf{u}$ preserves π, and L induces a linear transformation on π. This induced linear transformation on π preserves length and angles and so it is a rotation or a reflection on π.

Let L be a linear transformation from R^3 to R^3 that preserves length (and hence also angles). Then $L(\mathbf{u}) = P\mathbf{u}$ for some 3×3 orthogonal matrix P. We call L an **orthogonal transformation**. The geometric description of L depends on the eigenvalues of P.

Case 1. P has only one real eigenvalue, $k = 1$, and $P\mathbf{v} = \mathbf{v}$. Then L fixes a line in the direction of $\mathbf{v}$ and rotates the plane orthogonal to it.

Case 2. P has only one real eigenvalue, $k = -1$, and $P\mathbf{v} = -\mathbf{v}$. Then L is a rotation of the plane orthogonal to $\mathbf{v}$, followed by a reflection through that plane.

Case 3. P has three real eigenvalues, 1,1, and -1. Then L is a reflection through the plane containing the eigenvectors that go with $k = 1$.

Case 4. P has three real eigenvalues $-1,-1$, and 1. Then L fixes the line in the direction of the eigenvector going with 1, and rotates the plane orthogonal to it by 180 degrees.

Case 5. P has all real eigenvalues, $-1,-1$, and -1. L is a rotation by 180 degrees in the plane of two of them, followed by a reflection through this plane.

Case 6. If all three eigenvalues are 1, P is the identity.

Exercises

1. For each of the following orthogonal 2×2 matrices, determine whether they are reflections or rotations. If they are rotations, give the angle and if they are reflections, give the line through which the reflection takes place.

    (a) $\begin{bmatrix} \frac{3}{5} & \frac{4}{5} \\ -\frac{4}{5} & \frac{3}{5} \end{bmatrix}$

(b) $\begin{bmatrix} \frac{1}{\sqrt{2}} & -\frac{1}{\sqrt{2}} \\ -\frac{1}{\sqrt{2}} & -\frac{1}{\sqrt{2}} \end{bmatrix}$

(c) $\begin{bmatrix} \frac{\sqrt{3}}{2} & \frac{1}{2} \\ -\frac{1}{2} & \frac{\sqrt{3}}{2} \end{bmatrix}$.

2. Analyze the linear transformation from R^3 to R^3 associated with the orthogonal matrix

$$P = \begin{bmatrix} \frac{2}{7} & -\frac{6}{7} & \frac{3}{7} \\ \frac{3}{7} & -\frac{2}{7} & -\frac{6}{7} \\ \frac{6}{7} & \frac{3}{7} & \frac{2}{7} \end{bmatrix}$$

in terms of rotations and reflections.

3. Analyze the quadric surface represented by the quadric equation

$$x^2 + y^2 + 2z^2 - 2xy + 4xz + 4yz = 8,$$

by orthogonal diagonalization to standard form.

4. Same as problem 3 for the quadric surface given by

$$x^2 + y^2 + 2z^2 - 2xy - 4xz - 4yz = 16.$$

Appendix A

Change of Basis

Let $\{\mathbf{v}_1, \mathbf{v}_2, \cdots, \mathbf{v}_n\}$ be a basis of R^n. Let $\mathbf{w}$ be a vector in R^n. The $\mathbf{w}$ can be expressed uniquely as a linear combination of $\mathbf{v}_1, \mathbf{v}_2, \cdots, \mathbf{v}_n$, say

$$\mathbf{w} = c_1\mathbf{v}_1 + c_2\mathbf{v}_2 + \cdots + c_n\mathbf{v}_n.$$

The numbers $c_1, c_2, \cdots, c_n$ are called the coordinates of $\mathbf{w}$ with respect to the given basis. If we write $\mathbf{w}$ and $\mathbf{v}_1, \mathbf{v}_2, \cdots, \mathbf{v}_n$ as column vectors, then in matrix notation

$$\mathbf{w} = \begin{bmatrix} \mathbf{v}_1 & \mathbf{v}_2 & \cdots & \mathbf{v}_n \end{bmatrix} \begin{bmatrix} c_1 \\ c_2 \\ \vdots \\ c_n \end{bmatrix},$$

where

$$V = \begin{bmatrix} \mathbf{v}_1 & \mathbf{v}_2 & \cdots & \mathbf{v}_n \end{bmatrix}$$

is an $n \times n$ matrix with columns $\mathbf{v}_1, \mathbf{v}_2, \cdots, \mathbf{v}_n$. V is invertible since the columns of V are linearly independent, and we could solve for the coordinates of $\mathbf{w}$ using V^{-1}.

Let $\{\mathbf{u}_1, \mathbf{u}_2, \cdots, \mathbf{u}_n\}$ be another basis of R^n and suppose the two bases are related by the matrix equation

$$\begin{bmatrix} \mathbf{u}_1 & \mathbf{u}_2 & \cdots & \mathbf{u}_n \end{bmatrix} = \begin{bmatrix} \mathbf{v}_1 & \mathbf{v}_2 & \cdots & \mathbf{v}_n \end{bmatrix} P.$$

This equation simply says that column i of P is the coordinates of $\mathbf{u}_i$ with respect to the V basis. We know such a P exists (uniquely), because each $\mathbf{u}_i$ can be written uniquely as a linear combination of $\mathbf{v}_1, \mathbf{v}_2, \cdots, \mathbf{v}_n$. Given the two sets of basis vectors we could find P, since solving the above equation for P gives

$$P = \begin{bmatrix} \mathbf{v}_1 & \mathbf{v}_2 & \cdots & \mathbf{v}_n \end{bmatrix}^{-1} \begin{bmatrix} \mathbf{u}_1 & \mathbf{u}_2 & \cdots & \mathbf{u}_n \end{bmatrix}.$$

Remark *While P must be invertible since it is the product of two invertible matrices, P need not be orthogonal. However P is an orthogonal matrix if both of the bases involved are orthonormal bases.*

Remark *Notice that if the V basis is the standard basis, $v_1 = e_1$, $v_2 = e_2, \cdots, v_n = e_n$, then P is just the matrix whose columns are the U basis, $P = \begin{bmatrix} \mathbf{u}_1 & \mathbf{u}_2 & \cdots & \mathbf{u}_n \end{bmatrix}.$*

We see that if

$$\mathbf{w} = \begin{bmatrix} \mathbf{v}_1 & \mathbf{v}_2 & \cdots & \mathbf{v}_n \end{bmatrix} \begin{bmatrix} c_1 \\ c_2 \\ \vdots \\ c_n \end{bmatrix},$$

then

$$\mathbf{w} = \begin{bmatrix} \mathbf{u}_1 & \mathbf{u}_2 & \cdots & \mathbf{u}_n \end{bmatrix} P^{-1} \begin{bmatrix} c_1 \\ c_2 \\ \vdots \\ c_n \end{bmatrix},$$

which says that P^{-1} times the V coordinates of $\mathbf{w}$ is the U coordinates of $\mathbf{w}$.

Matrix of a Linear Transformation

Suppose $\{\mathbf{v}_1, \mathbf{v}_2, \cdots, \mathbf{v}_n\}$ and $\{\mathbf{u}_1, \mathbf{u}_2, \cdots, \mathbf{u}_n\}$ are two bases of R^n and are related by the matrix P as above. A linear transformation L is completely determined by its action on a basis. The matrix equation

$$\begin{bmatrix} L(\mathbf{v}_1) & L(\mathbf{v}_2) & \cdots & L(\mathbf{v}_n) \end{bmatrix} = \begin{bmatrix} \mathbf{v}_1 & \mathbf{v}_2 & \cdots & \mathbf{v}_n \end{bmatrix} A$$

Change of Basis

says that column i of A is the coordinates of $L(\mathbf{v}_i)$ with respect to the V basis.

We say that the matrix A represents L with respect to the V basis, or that A is the matrix of L with respect to the V basis.

Remark *If the V basis is the standard basis of R^n then A is called the **standard matrix** of L.*

If
$$\mathbf{w} = \begin{bmatrix} \mathbf{v}_1 & \mathbf{v}_2 & \cdots & \mathbf{v}_n \end{bmatrix} \begin{bmatrix} c_1 \\ c_2 \\ \vdots \\ c_n \end{bmatrix},$$

then
$$L(\mathbf{w}) = \begin{bmatrix} L(\mathbf{v}_1) & L(\mathbf{v}_2) & \cdots & L(\mathbf{v}_n) \end{bmatrix} \begin{bmatrix} c_1 \\ c_2 \\ \vdots \\ c_n \end{bmatrix}$$

$$= \begin{bmatrix} \mathbf{v}_1 & \mathbf{v}_2 & \cdots & \mathbf{v}_n \end{bmatrix} A \begin{bmatrix} c_1 \\ c_2 \\ \vdots \\ c_n \end{bmatrix}.$$

This says that A times the coordinates of $\mathbf{w}$ gives the coordinates of $L(\mathbf{w})$.

Suppose B is the matrix of L with respect to the basis $\{\mathbf{u}_1, \cdots, \mathbf{u}_n\}$. We wish to find out how A and B are related. We have

$$\begin{bmatrix} L(\mathbf{u}_1) & L(\mathbf{u}_2) & \cdots & L(\mathbf{u}_n) \end{bmatrix} = \begin{bmatrix} \mathbf{u}_1 & \mathbf{u}_2 & \cdots & \mathbf{u}_n \end{bmatrix} B. \quad \text{(A.1)}$$

Since
$$\begin{bmatrix} \mathbf{u}_1 & \mathbf{u}_2 & \cdots & \mathbf{u}_n \end{bmatrix} = \begin{bmatrix} \mathbf{v}_1 & \mathbf{v}_2 & \cdots & \mathbf{v}_n \end{bmatrix} P,$$

we have
$$\begin{bmatrix} L(\mathbf{u}_1) & L(\mathbf{u}_2) & \cdots & L(\mathbf{u}_n) \end{bmatrix} = \begin{bmatrix} L(\mathbf{v}_1) & L(\mathbf{v}_2) & \cdots & L(\mathbf{v}_n) \end{bmatrix} P.$$

Substituting the previous two equations in A.1 gives

$$\begin{bmatrix} L(\mathbf{v}_1) & L(\mathbf{v}_2) & \cdots & L(\mathbf{v}_n) \end{bmatrix} P = \begin{bmatrix} \mathbf{v}_1 & \mathbf{v}_2 & \cdots & \mathbf{v}_n \end{bmatrix} PB,$$

which gives

$$\begin{bmatrix} \mathbf{v}_1 & \mathbf{v}_2 & \cdots & \mathbf{v}_n \end{bmatrix} AP = \begin{bmatrix} \mathbf{v}_1 & \mathbf{v}_2 & \cdots & \mathbf{v}_n \end{bmatrix} PB.$$

Since $V = \begin{bmatrix} \mathbf{v}_1 & \mathbf{v}_2 & \cdots & \mathbf{v}_n \end{bmatrix}$ is invertible, we multiply both sides by V^{-1} to get

$$AP = PB,$$

or

$$P^{-1}AP = B.$$

Two $n \times n$ matrices A and B are called **similar** if there exists an invertible matrix P such that $P^{-1}AP = B$.

Fact *Two matrices represent the same linear transformation with respect to different bases if and only if the two matrices are similar.*

Reason. We have already shown that matrices representing the same linear transformation with respect to different bases are similar. To see that the converse is also true, assume that A and B are similar, or in other words that there exists an invertible matrix P such that

$$B = P^{-1}AP.$$

Let L be the linear transformation defined by

$$L(\mathbf{w}) = A\mathbf{w},$$

then A is the matrix of L with respect to the standard basis $\{\mathbf{e}_1, \cdots, \mathbf{e}_n\}$. Let the second basis $\{\mathbf{u}_1, \cdots, \mathbf{u}_n\}$ be the columns of P. Then

$$\begin{bmatrix} \mathbf{u}_1 & \mathbf{u}_2 & \cdots & \mathbf{u}_n \end{bmatrix} = \begin{bmatrix} \mathbf{e}_1 & \mathbf{e}_2 & \cdots & \mathbf{e}_n \end{bmatrix} P,$$

so by what we have shown above, the matrix of L with respect to $\{\mathbf{u}_1, \cdots, \mathbf{u}_n\}$ is $P^{-1}AP = B$. So A and B represent L with respect to different bases.

Change of Basis

Procedure *To find the standard matrix of a linear transformation L from R^n to R^n, find the images of the standard basis vectors of R^n, and put them as the columns of a matrix A.*

$$A = \begin{bmatrix} L(e_1) & L(e_2) & \cdots & L(e_n) \end{bmatrix}.$$

A is the standard matrix of L, and has the property that $L(u) = Au$ for all $u \in R^n$.

Remark *Given the standard matrix A of L, one can find the matrix B of L with respect to another basis $U = \{u_1, \cdots, u_n\}$, since they are related by the equation $B = P^{-1}AP$, where the columns of P are the vectors $u_1, \cdots, u_n$. This formula can also be solved for A.*

When $L(u) = Au$ and P is a matrix whose columns are a basis consisting entirely of eigenvalues of A and we form $P^{-1}AP$, we have found the matrix of L with respect to the basis of eigenvectors. This gives another proof that this must be a diagonal matrix, since if $\{u_1, \cdots, u_n\}$ is a basis of eigenvectors of L, then

$$\begin{bmatrix} L(u_1) & \cdots & L(u_n) \end{bmatrix} = \begin{bmatrix} k_1 u_1 & \cdots & k_n u_n \end{bmatrix}$$

$$= \begin{bmatrix} u_1 & \cdots & u_n \end{bmatrix} \begin{bmatrix} k_1 & 0 & \cdots & 0 \\ 0 & k_2 & & \vdots \\ \vdots & & \ddots & 0 \\ 0 & \cdots & 0 & k_n \end{bmatrix}.$$

Exercises

1. Show that if A and B are similar they have the same determinant, the same characteristic equation and the same eigenvalues.

2. Show that if $B = P^{-1}AP$ and if u is an eigenvector of A for the eigenvalue c, then $P^{-1}u$ is an eigenvector of B for c.

3. Let $\{v_1, v_2, v_3\}$ is a basis of R^3 and assume $L(v_1) = 2v_1 - 4v_2 + 7v_3$, $L(v_2) = v_1 + 3v_2 - 5v_3$, and $L(v_3) = 2v_3$. Find the matrix of L with respect to the basis $\{v_1, v_2, v_3\}$.

4. If $v_1 = (1, 4, 0)$, $v_2 = (2, 2, 1)$, and $v_3 = (1, -1, 2)$, and L is defined as in the previous problem, find the matrix of L with respect to the standard basis of R^3.

5. If T is a linear transformation from R^3 to R^3 defined by $T(x, y, z) = (3x - y, 2x + z, x + y + 2z)$, find the standard matrix A of T ($T(u) = Au$), and then find the matrix of T with respect to the basis in question 4.

6. Let $u_1 = (1, 2, 1)$, $u_2 = (2, 1, 1)$, and $u_3 = (1, -1, 1)$, and $\{v_1, v_2, v_3\}$ be as in problem 4. Find a matrix P relating these two bases.

7. If $w = (4, 8, 9)$, find the coordinates of w with respect to both bases from the preceding problem.

Appendix B

Answers to Selected Exercises

Section 1.1

1.
  (a) $\begin{bmatrix} 5 & 5 \\ 5 & 5 \end{bmatrix}$

  (b) $\begin{bmatrix} 0 & 0 \\ 0 & 0 \end{bmatrix}$

  (c) $\begin{bmatrix} 1 & 1 \\ 1 & 1 \end{bmatrix}$

  (d) $\begin{bmatrix} 1 & 0 & 0 \\ 0 & 4 & 0 \\ 3 & 0 & 3 \end{bmatrix}.$

2.
  (a) $\begin{bmatrix} 9 & 14 \\ -12 & 1 \end{bmatrix}$

  (b) $\begin{bmatrix} 9 & 14 & 4 \\ -12 & 1 & 5 \end{bmatrix}$

(c) $\begin{bmatrix} 1 & 6 & -2 \\ 12 & 7 & 11 \\ -8 & -22 & 2 \end{bmatrix}$

(d) undefined.

3.

(a) $\begin{bmatrix} 0 & 1 \\ 1 & 0 \end{bmatrix}$

(b) $\begin{bmatrix} 1 & 0 \\ 0 & 0 \end{bmatrix}$

(c) $\begin{bmatrix} 0 & 0 \\ 0 & 1 \end{bmatrix}$

(d) $\begin{bmatrix} 0 & 0 & 0 \\ 0 & 0 & 0 \\ 0 & 0 & 0 \end{bmatrix}$

(e) $\begin{bmatrix} 6 & 5 & 4 \\ 9 & 8 & 7 \\ 3 & 2 & 1 \end{bmatrix}$

(f) $\begin{bmatrix} 8 & 9 & 7 \\ 5 & 6 & 4 \\ 2 & 3 & 1 \end{bmatrix}$

(g) $\begin{bmatrix} 2 & 4 & 6 \\ 8 & 10 & 12 \\ 14 & 16 & 18 \end{bmatrix}$.

4.

(a) $\begin{bmatrix} c & 2c & 3c & 4c & 5c & 6c \end{bmatrix}$

(b) $\begin{bmatrix} c & 4d & 2c & 5d & 3c \end{bmatrix}$

(c) $\begin{bmatrix} e+5f & 2e+6f & 3e+7f & 4e+8f \end{bmatrix}$

(d) $[70]$

Answers to Selected Exercises

(e) $\begin{bmatrix} 5 & 10 & 15 & 20 \\ 6 & 12 & 18 & 24 \\ 7 & 14 & 21 & 28 \\ 8 & 16 & 24 & 32 \end{bmatrix}$

(f) $\begin{bmatrix} 6 & 6 & 6 \\ 1 & 4 & 1 \\ 1 & 1 & 4 \end{bmatrix}$

Section 1.2

1.
   (a) spans R^2
   (b) does not span R^2
   (c) spans R^2

2.
   (a) spans R^3
   (b) spans R^3
   (c) spans R^3
   (d) does not span R^3
   (e) does not span R^3

3.
   (a) not a basis
   (b) not a basis
   (c) is a basis.

4.
   (a) a basis
   (b) a basis
   (c) a basis

(d) not a basis

(e) not a basis

5.

(a) yes to both

(b) no to both

(c) no to both

(d) yes to both.

Section 1.3

1.

(a) $\begin{bmatrix} 1 & 0 \\ 0 & 1 \end{bmatrix}$

(b) $\begin{bmatrix} 1 & 0 & 0 \\ 0 & 1 & 0 \\ 0 & 0 & 1 \end{bmatrix}$

(c) $\begin{bmatrix} 1 & 0 & 0 & -1 \\ 0 & 1 & 0 & -1 \\ 0 & 0 & 1 & -1 \\ 0 & 0 & 0 & 0 \end{bmatrix}$

(d) $\begin{bmatrix} 1 & 2 & 0 \\ 0 & 0 & 1 \\ 0 & 0 & 0 \end{bmatrix}$

(e) $\begin{bmatrix} 1 & 0 & 0 & 0 \\ 0 & 1 & 0 & 0 \\ 0 & 0 & 1 & 0 \\ 0 & 0 & 0 & 1 \end{bmatrix}$

(f) $\begin{bmatrix} 1 & 0 & 3 & 2 \\ 0 & 1 & -1 & 4 \end{bmatrix}$

(g) $\begin{bmatrix} 1 & 0 \\ 0 & 1 \\ 0 & 0 \\ 0 & 0 \end{bmatrix}$.

Answers to Selected Exercises

2.
- (a) yes
- (b) yes
- (c) no
- (d) no
- (e) yes

3.
- (a) no
- (b) no
- (c) $x_1 = x_2 = x_3 = x_4 = 1$
- (d) $x_1 = 2$, $x_2 = -1$, $x_3 = 0$
- (e) no.

4. $x = 1$, $y = -2$, $z = 1$.

5. Column space basis: $\begin{bmatrix} 1 \\ 2 \\ 0 \\ 3 \end{bmatrix}$ and $\begin{bmatrix} 1 \\ 2 \\ 0 \\ 5 \end{bmatrix}$, row space basis: $(1, 2, 0, 1)$ and $(0, 0, 1, 1)$.

6.
- (a) $(1, 0, -5, -5)$ and $(0, 1, 2, 1)$.
- (b) $\begin{bmatrix} 1 \\ 1 \\ 2 \\ 3 \end{bmatrix}$ and $\begin{bmatrix} 3 \\ 4 \\ 3 \\ 8 \end{bmatrix}$.
- (c) $v_3 = 2v_2 - 5v_1$
- (d) $v_4 = v_2 - 5v_1$

7.

(a) $(1, 0, -2, 0)$, $(0, 1, 3, 0)$, $(0, 0, 0, 1)$.

(b) $\mathbf{u}_1$, $\mathbf{u}_2$, $\mathbf{u}_4$.

(c) $\mathbf{u}_3 = 3\mathbf{u}_2 - 2\mathbf{u}_1$.

(d) not possible.

8.

(a) dependent

(b) dependent

(c) independent

(d) independent

(e) independent.

Section 1.4

1.

(a) $\begin{bmatrix} -3 & 2 \\ 5 & -3 \end{bmatrix}$

(b) if $p^2 \neq 1$ it is invertible, and its inverse is

$$\begin{bmatrix} \frac{1}{1-p^2} & -\frac{p}{1-p^2} \\ -\frac{p}{1-p^2} & \frac{1}{1-p^2} \end{bmatrix}.$$

(c) not invertible

(d) not invertible

(e) $\begin{bmatrix} 6 & -8 & 3 \\ -\frac{7}{2} & 6 & -\frac{5}{2} \\ \frac{1}{2} & -1 & \frac{1}{2} \end{bmatrix}$

(f) $\begin{bmatrix} \frac{5}{18} & -\frac{1}{18} & -\frac{1}{18} \\ -\frac{1}{18} & \frac{5}{18} & -\frac{1}{18} \\ -\frac{1}{18} & -\frac{1}{18} & \frac{5}{18} \end{bmatrix}.$

2. $A = I$. Yes, again $A = I$.

5. See 1.(e). $x = 2$, $y = 1$, $z = 1$.

Answers to Selected Exercises

Section 1.5

1.
   - (a) 2
   - (b) 3
   - (c) 3
   - (d) 3
   - (e) 4

2. Here the answers are not unique (infinitely many ways to extend to a basis).

   - (a) $(1,0,0)$, or (using method of Sec. 1.6) $(\frac{1}{3}, \frac{1}{3}, -1)$
   - (b) $(1,0,0,0), (0,1,0,0)$, or $(1,0,-1,0), (0,1,0,-1)$
   - (c) $(1,0,0,0), (0,1,0,0)$, or $(-1,2,-1,0), (-2,3,0,-1)$.

4. $q \neq \frac{4}{7}$.

Section 1.6

1.
   - (a) $(-1, 2, -1)$
   - (b) $(1, 1, 1)$
   - (c) $(-3, 2, -1)$
   - (d) $(-4, 2, -1, 0)$
   - (e) $(1, 1, 1, 1)$.

2.
   - (a) Basis of row space: $(1, 0, -1)$ and $(0, 1, 2)$. Extend to basis of R^3 with $(-1, 2, -1)$.
   - (b) Basis of row space: $(1, 0, -1)$ and $(0, 1, -1)$. Extend to basis of R^3 with $(1, 1, 1)$.

(c) Basis of row space: $(1, 0, -3)$ and $(0, 1, 2)$. Extend to basis of R^3 with $(-3, 2, -1)$.

(d) Basis of row space: $(1, 0, -4, 0)$, $(0, 1, 2, 0)$ and $(0, 0, 0, 1)$. Extend to a basis of R^4 with $(-4, 2, -1, 0)$.

(e) Basis of row space: $(1, 0, 0, -1)$, $(0, 1, 0, -1)$ and $(0, 0, 1, -1)$. Extend to a basis of R^4 with $(1, 1, 1, 1)$.

3. If $A\mathbf{x} = \mathbf{O}$, then $A^{-1}(A\mathbf{x}) = A^{-1}\mathbf{O} = \mathbf{O}$, so $\mathbf{x} = \mathbf{O}$ is the only solution.

Section 2.1

1.

(a) 11
(b) 88
(c) 0
(d) $(x - y)^4$
(e) $(x^2 - 4x + 3)(x^2 - 5x + 4)$

2.

(a) I
(b) 4
(c) $\{\mathbf{O}\}$
(d) yes
(e) yes
(f) 176

3.

(a) is a basis
(b) is a basis
(c) is not a basis.

Answers to Selected Exercises 137

4. $k \neq -11$.

5. It does.

6. $-6e^{2x}$

7. $AA^{-1} = I$ so $|A|\,|A^{-1}| = |I| = 1$.

8. $|AB| = |A|\,|B|$ and $|B| = 0$ so $|AB| = 0$.

Section 2.2

1. $y = \frac{1}{2}x^2 + 3x + 2$.

2. $y = x^3 + 2x^2 + 3x + 4$.

3. $x = 2$.

4. $x = 2$, $y = 3$, $z = 4$.

5. $w = 1$.

6. $w = x = y = z = 1$.

Section 3.1

1.

(a) For $k = 5$, basis is $\begin{bmatrix} 1 \\ 1 \end{bmatrix}$, for $k = -1$, basis is $\begin{bmatrix} 1 \\ -1 \end{bmatrix}$.

(b) For $k = 5$, basis is $\begin{bmatrix} 1 \\ 1 \end{bmatrix}$, for $k = -2$, basis is $\begin{bmatrix} \frac{4}{3} \\ -1 \end{bmatrix}$.

(c) For $k = 1$, basis is $\begin{bmatrix} 2 \\ 2 \\ 1 \end{bmatrix}$, for $k = 5$, $\begin{bmatrix} 1 \\ 1 \\ 1 \end{bmatrix}$, for $k = -1$, $\begin{bmatrix} 1 \\ 2 \\ 1 \end{bmatrix}$.

(d) For $k=2$, $\begin{bmatrix} 1 \\ 1 \\ 0 \end{bmatrix}$ and $\begin{bmatrix} 0 \\ 0 \\ 1 \end{bmatrix}$.

(e) For $k=1$, $\begin{bmatrix} \frac{1}{2} \\ -1 \\ 0 \end{bmatrix}$, for $k=2$, $\begin{bmatrix} 0 \\ 0 \\ 1 \end{bmatrix}$.

(f) For $k=2$, $\begin{bmatrix} 1 \\ 1 \\ 0 \end{bmatrix}$ and $\begin{bmatrix} 1 \\ 0 \\ -1 \end{bmatrix}$, for $k=-1$, $\begin{bmatrix} \frac{1}{3} \\ -\frac{1}{3} \\ -1 \end{bmatrix}$.

(g) For $k=1$, $\begin{bmatrix} 1 \\ -1 \\ 0 \\ 0 \end{bmatrix}$, for $k=4$, $\begin{bmatrix} \frac{1}{2} \\ 1 \\ 0 \\ 0 \end{bmatrix}$, for $k=3$, $\begin{bmatrix} 0 \\ 0 \\ 1 \\ 1 \end{bmatrix}$, and

for $k=5$, $\begin{bmatrix} 0 \\ 0 \\ \frac{1}{2} \\ 1 \end{bmatrix}$.

2.

(a) $k=5$: 1 and 1. For $k=-1$: 1 and 1.

(b) $k=5$: 1 and 1. For $k=-2$: 1 and 1.

(c) $k=1$: 1 and 1. For $k=5$: 1 and 1. For $k=-1$: 1 and 1.

(d) $k=2$: 2 and 3.

(e) $k=1$: 1 and 2. For $k=2$: 1 and 1.

(f) $k=2$: 2 and 2. For $k=-1$: 1 and 1.

(g) 1 and 1 for all eigenvalues.

3. Use the fact that 0 is an eigenvalue of A if and only if $|A|=0$.

4. Same as preceding.

Answers to Selected Exercises

Section 3.2

1.
- (a) is diagonalizable.
- (b) is diagonalizable.
- (c) is diagonalizable.
- (d) is diagonalizable.
- (e) is diagonalizable
- (f) is not diagonalizable
- (g) is diagonalizable.
- (h) is diagonalizable.

2.

(a) $\begin{bmatrix} 3 & 1 \\ 5 & -1 \end{bmatrix}$

(c) $\begin{bmatrix} 1 & 1 \\ 1 & -1 \end{bmatrix}$

(e) $\begin{bmatrix} 2 & 1 & 1 \\ 1 & 0 & -2 \\ 0 & 1 & -1 \end{bmatrix}$

(g) $\begin{bmatrix} 1 & 1 & 0 & 0 \\ -1 & 1 & 0 & 0 \\ 0 & 0 & 1 & 0 \\ 0 & 0 & -1 & 1 \end{bmatrix}$

(h) $\begin{bmatrix} 1 & 1 & 1 & 1 \\ 1 & -1 & 0 & 0 \\ 1 & 0 & -1 & 0 \\ 1 & 0 & 0 & -1 \end{bmatrix}$.

Section 3.3

1.
   (a) $\begin{bmatrix} 2 \\ -1 \end{bmatrix}$

   (b) $\begin{bmatrix} 1 \\ 1 \\ 1 \end{bmatrix}$

   (c) $\begin{bmatrix} -1 \\ 2 \\ -1 \end{bmatrix}$

   (d) $\begin{bmatrix} \frac{1}{3} \\ \frac{1}{3} \\ 0 \\ 1 \end{bmatrix}$

   (e) $\begin{bmatrix} 2 \\ -1 \\ 0 \\ 0 \end{bmatrix}, \begin{bmatrix} 4 \\ 8 \\ 15 \\ -5 \end{bmatrix}$.

2.
   (a) $\begin{bmatrix} 1 \\ 2 \end{bmatrix}$

   (b) $\begin{bmatrix} 0 \\ 1 \\ -1 \end{bmatrix}, \begin{bmatrix} 2 \\ -1 \\ -1 \end{bmatrix}$

   (c) $\begin{bmatrix} 2 \\ 1 \\ 0 \end{bmatrix}, \begin{bmatrix} \frac{1}{5} \\ -\frac{2}{5} \\ -1 \end{bmatrix}$

   (d) $\begin{bmatrix} 1 \\ -1 \\ 0 \\ 0 \end{bmatrix}, \begin{bmatrix} 0 \\ 0 \\ 1 \\ 0 \end{bmatrix}, \begin{bmatrix} \frac{3}{2} \\ \frac{3}{2} \\ 0 \\ -1 \end{bmatrix}$

Answers to Selected Exercises

(e) $\begin{bmatrix} 3 \\ 6 \\ -4 \\ 0 \end{bmatrix}, \begin{bmatrix} 4 \\ 8 \\ 15 \\ 61 \end{bmatrix}$

3.

(a) $\frac{1}{\sqrt{5}} \begin{bmatrix} 2 \\ -1 \end{bmatrix}$

(b) $\frac{1}{\sqrt{3}} \begin{bmatrix} 1 \\ 1 \\ 1 \end{bmatrix}$

(c) $\frac{1}{\sqrt{6}} \begin{bmatrix} -1 \\ 2 \\ -1 \end{bmatrix}$

(d) $\frac{3}{\sqrt{11}} \begin{bmatrix} \frac{1}{3} \\ \frac{1}{3} \\ 0 \\ 1 \end{bmatrix}$

(e) $\frac{1}{\sqrt{5}} \begin{bmatrix} 2 \\ -1 \\ 0 \\ 0 \end{bmatrix}$

(f) $\frac{1}{\sqrt{330}} \begin{bmatrix} 4 \\ 8 \\ 15 \\ -5 \end{bmatrix}$.

Section 3.4

1.

(a) $\begin{bmatrix} \frac{1}{\sqrt{4-2\sqrt{2}}} & \frac{1}{\sqrt{4+2\sqrt{2}}} \\ \frac{\sqrt{2}-1}{\sqrt{4-2\sqrt{2}}} & \frac{-\sqrt{2}-1}{\sqrt{4+2\sqrt{2}}} \end{bmatrix}$

(b) $\begin{bmatrix} \frac{1}{\sqrt{2}} & \frac{1}{\sqrt{2}} \\ -\frac{1}{\sqrt{2}} & \frac{1}{\sqrt{2}} \end{bmatrix}$

(c) $\begin{bmatrix} \frac{1}{3} & \frac{2}{3} & \frac{2}{3} \\ \frac{2}{3} & \frac{1}{3} & -\frac{2}{3} \\ -\frac{2}{3} & \frac{2}{3} & -\frac{1}{3} \end{bmatrix}$

(d) $\begin{bmatrix} \frac{1}{3} & \frac{2}{3} & -\frac{2}{3} \\ \frac{2}{3} & \frac{1}{3} & \frac{2}{3} \\ -\frac{2}{3} & \frac{2}{3} & \frac{1}{3} \end{bmatrix}$

(e) $\begin{bmatrix} \frac{1}{\sqrt{3}} & \frac{1}{\sqrt{2}} & \frac{1}{\sqrt{6}} \\ \frac{1}{\sqrt{3}} & -\frac{1}{\sqrt{2}} & \frac{1}{\sqrt{6}} \\ \frac{1}{\sqrt{3}} & 0 & -\frac{2}{\sqrt{6}} \end{bmatrix}$

(f) $\begin{bmatrix} \frac{1}{2} & \frac{1}{2} & -\frac{1}{2} & -\frac{1}{2} \\ \frac{1}{2} & -\frac{1}{2} & -\frac{1}{2} & \frac{1}{2} \\ \frac{1}{2} & \frac{1}{2} & \frac{1}{2} & \frac{1}{2} \\ \frac{1}{2} & -\frac{1}{2} & \frac{1}{2} & -\frac{1}{2} \end{bmatrix}$

(g) $\begin{bmatrix} \frac{1}{\sqrt{2}} & 0 & \frac{1}{\sqrt{2}} & 0 \\ \frac{1}{\sqrt{2}} & 0 & -\frac{1}{\sqrt{2}} & 0 \\ 0 & \frac{1}{\sqrt{2}} & 0 & \frac{1}{\sqrt{2}} \\ 0 & \frac{1}{\sqrt{2}} & 0 & -\frac{1}{\sqrt{2}} \end{bmatrix}$.

3. $P^T A P = D$ implies $D^T = (P^T A P)^T = P^T A^T P$. Also $D^T = D$. From $P^T A P = P^T A^T P$, deduce that $A = A^T$.

Section 4.1

1. $x = -\frac{5}{9}$, $y = \frac{10}{3}$, $z = 4$.

2. $(3, 2, 5)$ is not in the span of the given vectors.

3. $\begin{bmatrix} -4 \\ -2 \\ 3 \\ -5 \end{bmatrix}$ is a basis of the solution space, thus $w = -4s$, $x = -2s$, $y = 3s$, $z = -5s$, gives all solutions.

4. $\begin{bmatrix} -1 \\ 3 \\ -1 \\ 0 \\ 0 \end{bmatrix}$ and $\begin{bmatrix} -16 \\ 5 \\ 0 \\ 5 \\ -1 \end{bmatrix}$ form a basis for the solution space.

Answers to Selected Exercises 143

5. $r \begin{bmatrix} -1 \\ 3 \\ -1 \\ 0 \\ 0 \end{bmatrix} + s \begin{bmatrix} -16 \\ 5 \\ 0 \\ 5 \\ -1 \end{bmatrix} + \begin{bmatrix} -6 \\ 2 \\ 0 \\ \frac{3}{2} \\ 0 \end{bmatrix}$.

Section 4.2

1. $\begin{array}{l} x_1(t) = 100e^{-4t} + 500 \\ x_2(t) = -50e^{-4t} + 250 \end{array}$. As $t \to \infty$, $x_1(t) \to 500$, $x_2(t) \to 250$. Stable in the long run. They will coexist.

2. $\begin{array}{l} x_1 = ae^t + be^{2t} + ce^{4t} \\ x_2 = ae^t + 2be^{2t} + 4ce^{4t} \\ x_3 = ae^t + 4be^{2t} + 16ce^{4t} \end{array}$.

3. $\begin{array}{l} x_1 = ae^{4t} + be^t + ce^t \\ x_2 = ae^{4t} - be^t \\ x_3 = ae^{4t} \qquad\quad - ce^t \end{array}$.

4. $\begin{array}{l} x_1 = 2ae^{8t} + be^{-t} \\ x_2 = ae^{8t} - 2be^{-t} - 2ce^{-t} \\ x_3 = 2ae^{8t} \qquad\quad + ce^{-t} \end{array}$.

Section 4.3

2.
$$A^q = \begin{bmatrix} \frac{2^q}{3} - \frac{2(-1)^{q+1}}{3} & \frac{2^q}{3} + \frac{(-1)^{q+1}}{3} \\ \frac{2^{q+1}}{3} - \frac{2(-1)^q}{3} & \frac{2^{q+1}}{3} + \frac{(-1)^q}{3} \end{bmatrix}.$$

3.
$$u_q = \frac{2^{q+1} + (-1)^{q+2}}{3}.$$

Section 4.4

1. $y = 2.9x + 3.2$

2. $x = \begin{bmatrix} \frac{16}{5} \\ \frac{5}{2} \\ -\frac{41}{4} \\ \frac{15}{4} \end{bmatrix}$.

Section 5.1

1. $x + 5y - z = 0$.

2. $x + 5y - z = 12$.

3. $x = 1$.

4. $(-2, 3, 2)$.

Section 5.2

1. $L_1 : 2, 1, 0.$ $L_2 : 1, 2, 5.$ $L_3 : -1, -1, -1.$

2. $\begin{aligned} x &= t + 1 \\ y &= 2t + 1 \\ z &= 3t + 1 \end{aligned}$.

3. $\begin{aligned} x &= t + 8 \\ y &= 3t + 7 \\ z &= 5t + 6 \end{aligned}$.

4. $\begin{aligned} x &= t + 6 \\ y &= -2t + 1 \\ z &= -3t + 1 \end{aligned}$.

5.
   (a) skew
   (b) parallel
   (c) intersecting
   (d) intersecting

Answers to Selected Exercises

Section 5.3

1. $\frac{3}{2}(1,1,2)$

2. $\frac{1}{2}(4,5,-3)$

3. 4

4. $\sqrt{14}$

5. $x + 2y + 3z = 34$

6.
   (a) both are $\frac{5}{\sqrt{6}}$

   (b) $\frac{10}{\sqrt{6}}$

   (c) yes

Section 5.4

1.
   (a) rotation. $\cos\tau = \frac{3}{5}$

   (b) reflection about the line through the vector $\left(\sqrt{2}+1,-1\right)$.

   (c) rotation. $\cos\tau = \frac{\sqrt{3}}{2}$, $\tau = 30°$.

2. Rotation of the plane $3x - y + 3z = 0$ about the line through the vector $(3,-1,3)$. This corresponds to Case 1.

3. Standard form is $2(x')^2 + 4(y')^2 - 2(z')^2 = 8$.

4. Standard form is $2(x')^2 + 4(y')^2 - 2(z')^2 = 16$.

Exercises at end of Appendix A

2. $BP^{-1}\mathbf{u} = P^{-1}APP^{-1}\mathbf{u} = P^{-1}A\mathbf{u} = P^{-1}c\mathbf{u} = cP^{-1}\mathbf{u}$.

3. $\begin{bmatrix} 2 & 1 & 0 \\ -4 & 3 & 0 \\ 7 & -5 & 2 \end{bmatrix}$.

4. Standard matrix of $L = \begin{bmatrix} 1 & 2 & 1 \\ 4 & 2 & -1 \\ 0 & 1 & 2 \end{bmatrix} \begin{bmatrix} 2 & 1 & 0 \\ -4 & 3 & 0 \\ 7 & 5 & 2 \end{bmatrix} \begin{bmatrix} 1 & 2 & 1 \\ 4 & 2 & -1 \\ 0 & 1 & 2 \end{bmatrix}^{-1}$

$= \begin{bmatrix} \frac{3}{7} & \frac{1}{7} & \frac{6}{7} \\ \frac{163}{7} & -\frac{53}{7} & -\frac{115}{7} \\ -\frac{122}{7} & \frac{48}{7} & \frac{99}{7} \end{bmatrix} = \frac{1}{7} \begin{bmatrix} 3 & 1 & 6 \\ 163 & -53 & -115 \\ -122 & 48 & 99 \end{bmatrix}$.

5. $A = \begin{bmatrix} 3 & -1 & 0 \\ 2 & 0 & 1 \\ 1 & 1 & 2 \end{bmatrix}$.

$B = \begin{bmatrix} 1 & 2 & 1 \\ 4 & 2 & -1 \\ 0 & 1 & 2 \end{bmatrix}^{-1} \begin{bmatrix} 3 & -1 & 0 \\ 2 & 0 & 1 \\ 1 & 1 & 2 \end{bmatrix} \begin{bmatrix} 1 & 2 & 1 \\ 4 & 2 & -1 \\ 0 & 1 & 2 \end{bmatrix}$

$= \frac{1}{7} \begin{bmatrix} 31 & 19 & 8 \\ -37 & -8 & 4 \\ 36 & 25 & 12 \end{bmatrix}$

6.

$P = \begin{bmatrix} 1 & 2 & 1 \\ 4 & 2 & -1 \\ 0 & 1 & 2 \end{bmatrix}^{-1} \begin{bmatrix} 1 & 2 & 1 \\ 2 & 1 & -1 \\ 1 & 1 & 1 \end{bmatrix}$

$= \frac{1}{7} \begin{bmatrix} 5 & -3 & -4 \\ -1 & 9 & 5 \\ 4 & -1 & 1 \end{bmatrix}$

relates the bases by

$\begin{bmatrix} \mathbf{u}_1 & \cdots & \mathbf{u}_n \end{bmatrix} = \begin{bmatrix} \mathbf{v}_1 & \cdots & \mathbf{v}_n \end{bmatrix} P$.

Answers to Selected Exercises

The inverse of P

$$P^{-1} = \frac{1}{3}\begin{bmatrix} 2 & 1 & 3 \\ 3 & 3 & -3 \\ 5 & -1 & 6 \end{bmatrix}$$

is also a correct answer since it relates the bases by

$$P^{-1}\begin{bmatrix} \mathbf{u}_1 & \cdots & \mathbf{u}_n \end{bmatrix} = \begin{bmatrix} \mathbf{v}_1 & \cdots & \mathbf{v}_n \end{bmatrix}.$$

7. Coordinates of **w** with respect to the U basis:

$$[\mathbf{w}]_U = \begin{bmatrix} 1 & 2 & 1 \\ 2 & 1 & -1 \\ 1 & 1 & 1 \end{bmatrix}^{-1} \begin{bmatrix} 4 \\ 8 \\ 9 \end{bmatrix} = \begin{bmatrix} 9 \\ -5 \\ 5 \end{bmatrix}.$$

Coordinates of **w** with respect to the V basis:

$$[\mathbf{w}]_V = \begin{bmatrix} 1 & 2 & 1 \\ 4 & 2 & -1 \\ 0 & 1 & 2 \end{bmatrix}^{-1} \begin{bmatrix} 4 \\ 8 \\ 9 \end{bmatrix} = \frac{1}{7}\begin{bmatrix} 40 \\ -29 \\ 46 \end{bmatrix}$$

Appendix C

MATS

About MATS

MATS is a computer program which will perform many matrix operations. It was written by Prof. Eugene Johnson, an experienced teacher of linear algebra. It was written especially for use by students taking a first course in linear algebra, and a couple of commands were added specifically to fit with this book. The big advantages of MATS are ease of use and the way it fits with the course. It allows the student to use the algorithms being taught while being relieved of the computational drudgery. The disk included with this book contains a copy of the MATS program designed to run on IBM compatible computers. If your computer has a hard drive, you can copy the file mats.exe from the disk onto your hard drive and run it from there. Also it is a good idea to make a copy of the disk supplied with this book.

Starting up MATS

To start the MATS program, put the disk in one of the drives (A or B) on your computer and change to the prompt for that drive (A:> or B:>). Now type *mats* and hit the Enter key. The program will load and you will see the title page with copyright information and a prompt > at the bottom left corner of the screen. You can now begin to use MATS. MATS is very easy to use. There is no complicated syntax to learn. To enter a matrix named A just type $mat(A)$ and hit the Enter key. After that the screen will prompt you what to do. By typing ? and hitting the Enter key, you will get a list of commands that MATS recognizes, and a brief discription of what they do. To exit the MATS program type *quit* and hit the Enter key. This is really all the instruction you need to start using MATS, but we include the following discussion of some of the commands and features which may be helpful.

Entering a matrix

To enter a matrix named A type $mat(A)$ and hit the Enter key. (Since it is necessary to hit the Enter key after every command we will not repeat this in what follows). The program will ask you for the number of rows. Enter an integer from 1 to 5. The program will then ask for the number of columns. Enter an integer from 1 to 10. The program will display a matrix of the designated size with zeros in all positions and with the cursor on the upper left position. Type in the number you want in this position and hit the Enter key. The number will appear in the first row first column position, and the cursor will move to the next entry in the first row. In this way you enter the matrix. You may enter fractions by typing 5/4, for example. If you type 10/8, MATS will respond 5/4. (MATS also has a decimal mode, but when the program starts up, it starts in fraction mode.) If you make a mistake in typing an entry for your matrix, you may correct with the backspace key before hitting the Enter key. Continue to enter the rest of the matrix. When the process is complete MATS will tell you "matrix stored as A". Now you can correct the matrix if necessary using the *edit* command. You can enter more matrices with different names by typing $mat(B)$, $mat(M)$, $mat(Q)$, $mat(ID)$, $mat(EXT)$, and so on. The name of a matrix may be one or more letters.

Remark $mat(B)$ *and mat B are the same to MATS. The parentheses are optional.*

Remark $mat(B)$ *and* $mat(b)$ *are not the same to MATS. MATS is case sensitive. The commands are written in lower case letters. It is recommended that you use upper case letters for naming matrices,* *especially if the name has more than one letter, to avoid confusion of names with commands.*

You can check on which names you have already used by typing *names*. MATS will return the names of all matrices currently defined. You can get rid of previously defined matrices by typing either *del(A)*, which deletes the matrix A, or *clr*, which deletes all previously defined matrices and frees up all the names.

Editing a Matrix

You can change individual entries in a matrix named A by typing *edit(A)*. MATS will display the matrix A, and ask which entry you wish to change. Enter the number of the row and then the number of the column containing the entry you wish to change. Then enter the number you wish to put in that position. When all your changes have been made, type q to quit editing. MATS will tell you "matrix A overwritten".

Some Useful Commands

Once you have entered your matrix, you can find its reduced row echelon form. Type $rref(A)$. MATS will return the RREF of the matrix A. Type $herm(A)$, and MATS will return the Hermite form of A. Type $ns(A)$ and MATS will return a matrix whose nonzero columns are a basis of the null space of A.

If A is a square matrix, $det(A)$ will return the determinant of A, $inv(A)$ will return the inverse of A, if A is invertible. The command $chp(A)$ will return the characteristic polynomial of A, and MATS will ask if you desire a rational root search (Y/N). If you select Y, MATS will find the rational roots and factor the characteristic polynomial. These rational roots are the rational eigenvalues. MATS will not find irrational or complex roots.

EROS

MATS has a subroutine called *eros* (standing for elementary row operations). This is very useful as and educational aid, because it allows you to do a step by step Gaussian elimination. Instead of just using the *rref* command, you can find the RREF by a step by step Gaussian elimination. If you have done the row reduction by hand, you can check each step on the computer. You can observe the algorithm at work without having to do all the arithmetic. If you type *eros*(A), MATS will display the matrix A and six operations you can perform.

1. < p > *piv*   If you type p, MATS will tell you it is going to perform "$k * \text{Row } I + \text{Row } J$", and ask you to enter first k, then I and J. (The *piv* stands for pivot.)

2. < f > *fpiv*   If you type f, MATS will tell you it is going to "*Pivot on the* $(I,J)-entry$", and will ask you to enter I and then to enter J. MATS will add multiples of row I to all other rows so that column J will be zero except in row I. This operation is called a full pivot. Of course it can't be done if the (I,J) entry of your matrix is 0.

3. < s > *swap*   If you type s, MATS will reply "Row $I \longleftrightarrow$ Row J", and ask you to enter I and then J. MATS will exchange row I and row J.

4. < $*$ > *rowp*   If you type $*$, MATS will reply "$k*\text{Row } I, \text{Enter } k$". Then it will ask you to enter I. MATS will multiply row I by k.

5. < / > *rowd*   If you type / , MATS will reply "$1/k * \text{Row } I$", and ask you to enter first k and then I. MATS will divide row I by k.

6. < q > *quit*   If you type q , MATS will exit the *eros* subroutine.

Remark   *If you wish to name the last matrix that has appeared B, just type $= B$ at the prompt. MATS will reply "matrix stored as B". Next time you type B, MATS will display the matrix you have named B.*

The Record and Remark Commands

If you type *record* (followed by Enter), MATS will tell you "the recorder is now off, would you like to change it?(Y/N)". If you choose Y, MATS will ask for a file name. Enter a name for the file. Everything that appears on the screen after that (until the recorder is turned off) will be recorded in this file. After you have exited the MATS program, you can print out this file with the DOS print command. The file can also be edited and printed with any standard word processing program.

If you type *rem*, this tells MATS that you want to insert something that is not a command, but that will appear when the file is typed out. This command is useful if you want to insert a remark into some computations that you are recording and plan to print out.

List of Commands

Remark *Parentheses are optional and can be replaced by a space.*

$mat(A)$: calls matrix collection routine for matrix named A.

$edit(A)$: calls matrix edit routine to edit matrix A.

$det(A)$: returns the determinant of A.

$inv(A)$: returns the inverse of A if A is invertible.

$rref(A)$: returns the reduced row echelon form of A.

$herm(A)$: returns the Hermite form of A.

$ns(A)$: computes $I - Herm(A)$, the nonzero columns of which are a basis for the null space of A.

$chp(A)$: returns the characteristic polynomial of A.

$adj(A)$: returns the adjoint of A.

$rowmat(A)$: calls collection routine for a row matrix named A.

$rmat(A)$: abbreviation for rowmat(A).

$colmat(A)$: calls collection routine for a column matrix named A.

$cmat(A)$: abbreviation for colmat(A).

imat(A): generates an identity matrix named A (prompts for size.)

aug(A, B): returns the augmented matrix $[A\ B]$.

eros(A): calls the elementary row operations routine on A.

piv(A): calls the pivot routine on A.

fpiv(A): calls the full pivot routine on A.

rowp(A): calls the row multiplication routine on A.

rowd(A): calls the row division routine on A.

swap(A): calls the row swap routine on A.

$A + B$: adds matrices A and B.

$A - B$: subtracts B from A.

$A * B$: multiplies A and B.

$<\mathbf{u}, \mathbf{v}>$: returns the dot product (Euclidean Inner Product) of $\mathbf{u}$ and $\mathbf{v}$.

$\mathbf{u} \times \mathbf{v}$: returns the cross product of $\mathbf{u}$ and $\mathbf{v}$.

ca(m, n, A): returns columns m through n of A as the new active matrix.

row(i, A): returns row i of A.

col(i, A): returns column i of A.

names: lists names of matrices defined.

$= A$: names the current contents of the Xregister as A.

clr: clears all previously used matrices and frees memory.

del(A): deletes the matrix A and frees memory.

mode: reports mode (fract/dec) and allows toggle.

$cof(i, j, A)$: returns the (i,j)-cofactor of A.

$cofmat(A)$: returns the matrix of cofactors of A.

$glinv(A)$: returns an invertible matrix P such that $P * A = RREF(A)$.

A/B: returns $A*\text{glinv}(B)$.

$B\backslash A$: returns $\text{glinv}(B)*A$.

xreg: returns the current contents of the Xregister.

dec: switches to decimal mode.

fract: switches to fraction mode.

fix: sets number of decimal places dispalyed in decimal mode.

record: reports recorder status (on/off) and allows for toggle.

rem: tells MATS to ignore current line except to record it if recorder is on.

? : brings up a list of commands and brief discription of what they do.

quit: exits MATS, returns you to DOS prompt.

Remark *If you type a command without an argument, the command will be applied to the last matrix the program has seen (current contents of the Xregister).*

Warning *If you want to install MATS on a network, contact the author for permission and further instructions.*

Index

Addition, Vector, 8
Algebraic Multiplicity, 56
Arithmetic, Matrix, 3-6
Characteristic Equation, 53
Characteristic Polynomial, 53
Complement, Orthogonal, 72
Cramer's Rule, 47
Cross Product and Planes in R^3, 103-106
Determinant, Vandermonde, 48
Determinants, 39-50
Determinants, Application of, 47-50
Diagonal Matrix, 59
Diagonalization, 39-50
Diagonalization, Orthogonal, 73-82
Differential Equations, 91-95
Dimension, Rank and, 27-32
Direction Numbers, 108
Echelon Form, Reduced Row, 12-20
Eigenvalues and Eigenvectors, 51-62
Elementary Matrices, 24
Elementary Row Operations, 13, 18
Expansion by Minor, 43-45
Fibonacci Sequence, Powers of a Matrix and the , 96-98
Geometric Multiplicity, 56
Geometry, 103-122
Gram-Schmidt, 81
Hermite Form, Null Space and, 33-38
Idempotent, 27
Inverse, 21-26
Kirchoff's Law, 88
Least Squares Approximation, 99-102
Linear Equations, 83-90
Linear Transformation, 51

Lines in R^3, 107-112
Matrices and Quadrics, Orthogonal 115-122
Matrices, Elementary, 24
Matrices, Vandermonde, 49
Matrices, Vectors and, 3-38
Matrix and the Fibonacci Sequence, Powers of a , 96-98
Matrix Arithmetic, 3-6
Matrix, Diagonal, 59
Matrix, Standard, 125
MATS, 149-155
multiplication, Scalar, 8
Multiplicity, Algebraic, 56
Multiplicity, Geometric, 56
Null Space and Hermite Form, 33-38
Orthogonal Basis, 66-72
Orthogonal Complement, 72
Orthogonal Diagonalization, 73-82
Orthogonal Matrices and Quadrics 115-122
Parametric Equations, 107
Planes in R^3, Cross Product and, 103-106
Polynomial, Characteristic, 53
Powers of a Matrix and the Fibonacci Sequence, 96-98
Quadric Surface, 119
Quadrics, Orthogonal Matrices and 115-122
R^3, Lines in, 107-112
Rank and Dimension, 27-32
Reduced Row Echelon Form, 12-20
Rotation, 118
Row Operations, Elementary, 13, 18

INDEX

Scalar multiplication, 8
Skew, 109
Standard Matrix, 125
Unit Vectors, 72
Vandermonde Determinant, 48

Vandermonde Matrices, 49
Vector addition, 8
Vectors and Matrices, 3-38
Vectors, 7-11
Vectors, Unit, 72